ISOTOPES
IN ORGANIC CHEMISTRY

VOLUME 5
ISOTOPES IN CATIONIC
REACTIONS

ISOTOPES IN ORGANIC CHEMISTRY

Edited by

E. BUNCEL

Queen's University, Kingston, Ontario, Canada

and

C.C. LEE

University of Saskatchewan, Saskatoon, Saskatchewan, Canada

VOLUME 5
ISOTOPES IN CATIONIC
REACTIONS

ELSEVIER SCIENTIFIC PUBLISHING COMPANY
Amsterdam—Oxford—New York 1980

ELSEVIER SCIENTIFIC PUBLISHING COMPANY
335 Jan van Galenstraat
P.O. Box 211, 1000 AE Amsterdam, The Netherlands

Distributors for the United States and Canada:

ELSEVIER NORTH-HOLLAND INC.
52, Vanderbilt Avenue
New York, N.Y. 10017

Library of Congress Cataloging in Publication Data
Main entry under title:

Isotopes in cationic reactions.

   (Isotopes in organic chemistry ; v. 5)
   Includes bibliographical references and index.
   1. Chemical reactions. 2. Cations. 3. Isotopes.
I. Buncel, E.   II. Lee, Choi Chuck, 1924–
III. Series.
QD501.I818   541.3'9   80-23222
ISBN 0-444-41927-6

ISBN 0-444-41927-6 (Vol. 5)
ISBN 0-444-41742-7 (Series)

Printed in The Netherlands

ISOTOPES IN ORGANIC CHEMISTRY

ADVISORY BOARD

Contributors to Volume 5

E. Buncel

Department of Chemistry
Queen's University
Kingston, Ontario K7L 3N6 CANADA

T.L. Gibson

Department of Chemistry
University of Texas at Austin
Austin, Texas 78712 U.S.A.

C.C. Lee

Department of Chemistry and Chemical
    Engineering
University of Saskatchewan
Saskatoon, Saskatchewan S7N 0W0 CANADA

T. Numata

Department of Chemistry
University of Tsukuba
Sakura-mura, Ibaraki, 300-31 JAPAN

S. Oae

Department of Chemistry
University of Tsukuba
Sakura-mura, Ibaraki, 300-31 JAPAN

R.M. Roberts

Department of Chemistry
University of Texas at Austin
Austin, Texas 78712 U.S.A.

D.L.H. Williams

Department of Chemistry
University of Durham
Durham, ENGLAND

ISOTOPES IN ORGANIC CHEMISTRY

Volume 1. Isotopes in molecular rearrangements

N.C. Deno
The Pennsylvania State University

Deuterium labeling in carbonium
ion rearrangements

W.R. Dolbier, Jr.
University of Florida

Isotope effects in pericyclic
reactions

J. L. Holmes
University of Ottawa

The elucidation of mass spectral
fragmentation mechanisms by
isotopic labeling

D. H. Hunter
University of Western Ontario

Isotopes in carbanion rearrangements

J.S. Swenton
The Ohio State University

Utilization of deuterium labeling
in organic photochemical rearrangements

Volume 2. Isotopes in hydrogen transfer processes

M.M. Kreevoy
University of Minnesota

The effect of structure on isotope
effects in proton transfer
reactions

G. Lamaty
Université de Montpellier

Deuterium exchange in carbonyl
compounds

K.T. Leffek
Dalhousie University

Proton transfers in nitro compounds

E.S. Lewis
Rice University

Isotope effects in hydrogen atom
transfer reactions

H. Simon and A. Kraus
Technische Universität Munich

Hydrogen isotope transfer in
biological processes

P.J. Smith
University of Saskatchewan

Isotope effects in elimination
reactions

R. Stewart
University of British Columbia

Isotopes in oxidation reactions

Volume 3. Carbon-13 in organic chemistry

G.E. Dunn University of Manitoba	Carbon-13 kinetic isotope effects in decarboxylation
J. Hinton, M. Oda and A. Fry University of Arkansas	Carbon-13 n.m.r. methodology and mechanistic applications
G. Kunesch and C. Poupat Institute de Chimie des Substances Naturelles du C.N.R.S.	Biosynthetic studies using carbon-13 enriched precursors
A.S. Perlin McGill University	Application of carbon-13 n.m.r. to problems of stereochemistry
A.V. Willi Universitat Hamburg	Kinetic carbon and other isotope effects in cleavage and formation of bonds to carbon

Volume 4. Tritium in Organic Chemistry

J.A. Elvidge and J.R. Jones University of Surrey V.M.A. Chambers and E.A. Evans Radiochemical Centre, Amersham	Tritium nuclear magnetic resonance spectroscopy
W.J. Spillane University College, Galway	The use of tritium and deuterium in photochemical electrophilic aromatic substitution
Y.-N. Tang Texas A & M University	Reactions of energetic tritium atoms with organic compounds
D.W. Young University of Sussex	Stereospecific synthesis of tritium labelled organic compounds using chemical and biological methods

FOREWORD TO VOLUME 1

   Organic chemistry is characterized by a vast variety of compounds, structures and reactions realized by a rather limited number of chemical elements. One and the same element is generally represented by a considerable number of atoms, playing several different roles. It is evident that a method enabling us to give the otherwise anonymous atom a kind of identity should be of particular value in this branch of chemistry.

   Tracing by means of similar but still chemically discernible groups has been practised in organic chemistry for a long time, and has revealed that organic reactions are far more varied than expected. Isotopes, being chemically identical in a qualitative and usually also in an almost quantitative sense, are far more powerful as tracers, due to this similarity and to the fact that atoms rather than groups are labelled and can be traced as such.

   A simple account of the molecular species involved, their structures and configurations, cannot be considered a complete description of a chemical system in equilibrium. We know from studies of non-equilibrium systems that opposite reactions balancing one another are generally taking place. As far as species of different molecular compositions are concerned, this has been realized for more than a century. It is only in the last few decades, however, that we have had at hand the means to measure the amounts of different isotopes and follow the behaviour of systems which are non-equilibrium ones with respect to isotopic composition. This has led to a still more vivid picture of most systems in equilibrium, with several exchange reactions taking place, sometimes at rates too large to be measured on the classical time scale of chemical reactions.

   Even this is not enough for the true scientist who wants to go beyond the knowledge of which reactions actually take place and how fast they occur. He feels a desire to know also how the atomic nuclei and electrons behave in the transition called a chemical reaction. His questions come close to the fundamental limit set by the principle of uncertainty. At present the transition state of the rate-determining reaction step seems to be the most complete description attainable. In such studies it is not only the qualitative chemical similarity of isotopes, allowing the identity of atoms in the transition state to be revealed, but also their quantitative chemical dissimilarity which is of importance and allows a study of the force field and hence binding conditions in the transition state itself. Thanks to the fairly low atomic number of most atomic species of importance in organic chemistry, the relative mass differences between isotopes are sufficient to cause differences in quantitative behaviour, rather easily measurable with modern instruments.

   Many scientists feel the flood of scientific publications as an encumbrance. The justification for the existence of a series like the one started by the present volume lies in the aid that the surveys it contains may offer the research worker and, perhaps more important, the stimulus for further research that may be provided. The application of isotope methods has undoubtedly a very important role in future research in organic chemistry. No attempt at a detailed prediction will be ventured here, however. It may suffice to refer to the development of the nuclear magnetic resonance technique. The studies of ordinary hydrogen nuclei, which have been of outstanding importance for the development of organic chemistry, can be considered as an application of isotope methods according to the ordinary usage of the concept only to the extent that deuterium has been used as a stand-in for protium. In the not-too-distant future, however, most laboratories will have equipment allowing routine studies of the less abundant carbon isotope ^{13}C, and then many chemists will be in possession of a sensitive probe in the centre of atoms of the most important element in organic chemistry. It will reveal not only details of molecular structure

in the usual sense but also more subtle details about the electron
distribution in the backbone of organic molecules. It is open to discussion,
of course, whether this kind of work, which frequently makes use of the
natural occurrence of heavy carbon, should be considered as an application
of isotope methods. In any case, it utilizes a particular property of an
isotope different from the most abundant one.

It is evident from the thoughts expressed in the last paragraph that
the borders of the field "Isotopes in Organic Chemistry" are rather
indeterminate. The editors' intention to apply as few restrictions as
possible on subject matter seems wise, because then the interest taken in
the present series by its future readers can be allowed to indicate the
position of these borders in practice.

Göteborg   Lars Melander

Contents

Chapter 1

DEGENERATE REARRANGEMENTS IN TRIARYLVINYL CATIONS

C. C. LEE

Department of Chemistry and Chemical Engineering, University of Saskatchewan, Saskatoon, Saskatchewan, Canada, S7N 0W0

I. INTRODUCTION

Prior to the work of Grob and Cseh,[1] vinyl cations were generally regarded as unattractive reaction intermediates. This belief arose largely because simple vinyl halides are inert towards reagents such as alcoholic silver nitrate.[2,3] In 1964, Grob and Cseh[1] reported that α-bromostyrenes la-d undergo unimolecular solvolysis

$$\underline{p}-X-C_6H_4-\underset{\underset{\underline{1}}{Br}}{\overset{}{C}}=CH_2 \qquad \underline{a} \quad \underline{b} \quad \underline{c} \quad \underline{d} \qquad \underline{e}$$

$$X = H \quad NH_2 \quad CH_3O \quad CH_3CONH \quad NO_2$$

in 80% EtOH to give the corresponding acetophenones as products. The relative rates at 100° for the p-amino-, p-methoxy- and p-acetamido-derivatives lb-d were about 10^8, 10^4 and 10^3 times, respectively, as fast as the unsubstituted parent la. The rates were insensitive to the presence of 1-5 equivalents of Et_3N, and with la and lc, there was catalysis by Ag^+ ion. In contrast, le reacted only in the presence of Et_3N via a bimolecular elimination to give p-nitrophenylacetylene. These results led to the proposal that α-aryl substituted vinyl cations were formed in the solvolysis of la-d. Since this first demonstration, from chemical and kinetic evidence, of vinyl cations as reaction intermediates, a great deal of work on vinyl cations has been done and a number of reviews have been written on the subject.[4-9]

Early applications of isotopic tracers in the study of vinyl cations were made by Modena and coworkers.[10,11] In the solvolysis of trans-1,2-diaryl-2-arylthiovinyl 2,4,6-trinitrobenzenesulfonates (2), the common-ion effect was confirmed by the use of ^{35}S-labelled lithium 2,4,6-trinitrobenzenesulfonate (LiTNBS).

Ar   OSO₂R
   \   /
    C=C
   /   \
ArS   Ar

$$\text{Ar}\diagdown \underset{\diagup}{\overset{\diagdown}{C}}=\underset{\diagup}{\overset{\diagdown}{C}}\diagdown \text{Ar}$$

(structures)

Ar   OSO₂R   Ar   Ar
  \   /   \   /
   C = C   C = C
  /   \   \+   /
ArS   Ar   S
                                      |
                                      Ar

          2   3

After partial reaction in the presence of the labeled LiTNBS,
radioactivity was shown to be incorporated into the unreacted
vinyl sulfonate.[10] It was also shown that the solvolysis of
trans-1,2-diphenyl-2-phenylthio[1-^{14}C]vinyl brosylate* gave a
product derived from the thiirenium cation 3, resulting in the
complete scrambling of the label between the two ethylenic carbons.[11]
These and related studies have been reviewed by Modena and Tonellato,[6]
and by Stang.[7]

   Vinyl cations undergo reactions typical of carbocations including
rearrangements.[4-9] An early example of a 1,2-aryl shift in a
triarylvinyl cation was reported by Jones and Miller[12] in their
study on the generation of vinyl cations from vinyltriazenes.   Thus
the decomposition of 1-phenyl-2,2-di-p-tolylvinylphenyltriazene (4)
in HOAc gave 80% and 20%, respectively, of the unrearranged and
rearranged acetates 5 and 6. Similarly, the decomposition of

Tol   NHNNPh   Tol   OAc   Tol   OAc
  \   /   \   /   \   /
   C = C   C = C   C = C
  /   \   /   \   /   \
Tol   Ph   Tol   Ph   Ph   Tol

      4   5   6

1-methyl-2,2-diphenylvinylphenyltriazene (7a) in HOAc gave largely
the rearranged acetate 8. The first example of a carbon migration

Ph   X   Ph   OAc   Ph   O
  \   /   \   /   \   ‖
   C = C   C = C   CH-C-Ph
  /   \   /   \   /
Ph   CH₃   CH₃   Ph   CH₃

7a X = NHNNPh   8   9
7b X = OTf

---

*The naming of labelled compounds in this chapter will follow IUPAC
provisional rules given in IUPAC Information Bulletin No. 62, July, 1977
Nomenclature of Organic Chemistry, Section H: Isotopically Modified Compound

across the double bond of a vinyl cation generated by solvolysis
was reported in 1970 by Schleyer, Hanack, Stang and their coworkers.[13]
When l-methyl-2,2-diphenylvinyl triflate (7b) was solvolyzed in 80%
EtOH, followed by ether hydrolysis, the product was almost exclusively
the rearranged ketone 9.

In the above examples, a less stable vinyl cation is rearranged to
a more stable one. In degenerate rearrangements, both the initial
and rearranged ions have the same energy and the same chemical
structure, and such processes are detected through the use of isotopic
labels. A considerable amount of work on degenerate rearrangements
in triarylvinyl cations have been done in recent years. In keeping
with the common theme of this series in highlighting the use and
value of isotopes in organic chemistry, degenerate rearrangements
arising from 1,2-aryl shift across the double bond in triarylvinyl
cations, as illustrated in eqs. 1 and 2, are reviewed in this chapter.

$$
\begin{matrix} Ar & & & & Ar \\ & \diagdown & + & & + & \diagup \\ & & C^*{=}C{-}Ar & \rightleftharpoons & Ar{-}C^*{=}C & \\ & \diagup & & & & \diagdown \\ Ar & & & & & Ar \end{matrix} \qquad (1)
$$

$$
\begin{matrix} Ar' & & & & Ar' \\ & \diagdown & + & & + & \diagup \\ & & C^*{=}C{-}Ar & \rightleftharpoons & Ar{-}C^*{=}C & \\ & \diagup & & & & \diagdown \\ Ar & & & & & Ar \end{matrix} \qquad (2)
$$

II. THE TRIPHENYLVINYL SYSTEM

In their work on the generation of vinyl cations from vinyltriazenes,
Jones and Miller[12] reported in 1967 the formation, among other vinyl
cations, of the triphenylvinyl cation from the decomposition of
triphenylvinylphenyltriazene (10a); e.g., 10a decomposes in HOAc
to give triphenylvinyl acetate (10b) in quantitative yield. In the

$$
\begin{matrix} Ph & & X \\ & \diagdown & \diagup \\ & & C{=}C \\ & \diagup & \diagdown \\ Ph & & Ph \\ & 10 & \end{matrix}
$$

X=NNNHPh; OAc; I; OSO_2F; OTf; OTs; Br
  a   b   c   d   e   f   g

following year, Miller and Kaufman[14] showed that the solvolysis, in
aqueous dimethylformamide, of a number of triarylvinyl iodides,
including triphenylvinyl iodide (10c), proceeded via an S_N1 type
of mechanism. The solvolysis of triphenylvinyl fluorosulfonate
(10d), trifluoromethanesulfonate(triflate) (10e), or p-toluene-
sulfonate(tosylate) (10f) in HOAc or EtOH-H_2O was investigated

by Jones and Maness,[15] and both product and kinetic evidence pointed to a simple S_N1 mechanism via the triphenylvinyl cation as reaction intermediate.

In the work of Jones and Maness,[15] some applications of isotopes were also utilized. Thus the absence of a solvent kinetic isotope effect when 10d or 10f was solvolyzed in HOAc or DOAc was taken as evidence against an addition-elimination mechanism. Moreover, a secondary β-deuterium isotope effect, k_H/k_D, of 1.45 was observed in the acetolysis of 1-phenyl(2-^{2}H)vinyl fluorosulfonate, HDC=C(Ph)-OSO$_2$F. This was regarded as positive support for the S_N1 heterolytic cleavage, since this value was in good agreement with secondary β-deuterium isotope effects reported for carbocationic processes in both saturated and unsaturated systems.[16-18]

Extensive studies on the solvolysis, in 60% EtOH and in 2,2,2-trifluoroethanol (TFE), of triphenylvinyl bromide (10g) as well as cis- and trans-2-anisyl-1,2-diphenylvinyl and 2,2-dianisyl-1-phenylvinyl bromides† (cis- and trans-11 and 12, respectively)

An   X   X = Br OTFE OAc   An   Br
  \   /   \   /
   C=C   a   b   c   C=C
  /   \   /   \
Ph   Ph   An   Ph

    11   12

have been carried out in 1973 by Rappoport and Houminer.[19] From the kinetic data and the nature of the products, including rearrangement products, from cis- and trans-11 and from 12, it was concluded that these triarylvinyl bromides solvolyzed via initial ionization without any β-aryl participation to give the open, linear vinyl cation, which could return to covalent bromide, rearrange, and react with solvent or with added nucleophile to give the products.

A. Solvolysis of triphenyl[2-^{14}C]vinyl triflate (10e-2-^{14}C)

The first example of a degenerate rearrangement arising from a 1,2-aryl shift across the double bond in a triarylvinyl cation was reported from this laboratory in 1974.[20] Solvolyses of triphenyl-[2-^{14}C]vinyl triflate (10e-2-^{14}C) were carried out in glacial HOAc,

---

†In accordance with the usage adopted in a number of published papers on triarylvinyl cations, anisyl, rather than p-anisyl, is used to designate the p-methoxyphenyl group.

97% HCOOH and CF_3COOH. It was found that the products showed scramblings of the label from C-2 to C-1 averaging 6.7, 7.7 and 27.0%, respectively, for the acetolysis, formolysis and trifluoro-acetolysis, as illustrated, for example, by eq. 3. Interestingly,

$$
\begin{array}{ccc}
\underset{Ph}{\overset{Ph}{\diagdown}}\overset{14}{C}=C\overset{\diagup OTf}{\diagdown Ph} & \xrightarrow{\text{HOAc}} & \underset{Ph}{\overset{Ph}{\diagdown}}\overset{14}{C}=C\overset{\diagup OAc}{\diagdown Ph} + \underset{Ph}{\overset{AcO}{\diagdown}}\overset{14}{C}=C\overset{\diagup Ph}{\diagdown Ph}
\end{array} \quad (3)
$$

$\underline{10e}\text{-}2\text{-}^{14}C$ $\underline{10b}\text{-}2\text{-}^{14}C$ $\underline{10b}\text{-}1\text{-}^{14}C$

(93.3%)   (6.7%)

the presence of 1.1 equiv. of the conjugate base of the solvent, added as the sodium salt, did not materially affect the extent of scrambling. This finding was in contrast with the observation of Jones and Miller,[12] who found that in the decomposition of triazene **4** in HOAc, the formation of the 20% rearranged product **6** was completely suppressed by the presence of 10 equiv. of KOAc.

Since the decomposition of a vinyltriazene presumably proceeds via the diazonium ion, which gives rise to a "free" vinyl cation that, in turn, may competitively undergo rearrangement or be captured by a nucleophile, the presence of an added stronger nucleophile, such as the conjugate base of the solvent, is expected to decrease the extent of rearrangement as was observed by Jones and Miller. As the presence of the conjugate base did not affect the extent of scrambling in the solvolysis of $\underline{10e}\text{-}2\text{-}^{14}C$,[20] it was suggested that the 1,2-phenyl shift in the triphenylvinyl cationic system may have occurred during the ion-pair stage. An analogy involving ion-pairs can be found in the classical work of Young, Winstein and Goering[21,22] on allylic rearrangements, through internal returns with allylic chlorides, in which intimate ion-pairs were involved and the presence of added chloride ion did not affect the rate of rearrangement.

Further work in support of an important role for ion-pairs in the triphenylvinyl cationic system will be discussed later in Section II. B.2. However, it may be of interest to point out here that the decomposition of triphenyl[2-^{14}C]vinylphenyltriazene ($\underline{10a}\text{-}2\text{-}^{14}C$) in a number of acids did not give rise to any isotopically scrambled product.[23] Possibly, a "hot" triphenylvinyl cation generated in this way may have too short a lifetime for the 1,2-phenyl shift to compete with product formation, while under solvolytic conditions, the life-time of the ion-pair may be sufficiently long to allow some 1,2-

phenyl shifts to take place.

In our work with 10e-2-^{14}C,[20] the 6.7, 7.7 and 27.0% scramblings, respectively, found in the acetolysis, formolysis and trifluoro-acetolysis, have been contrasted with the approximately 6, 45 and 50% scramblings, respectively, observed in the solvolysis of 2-phenylethyl tosylate (13a), labelled at C-1 with ^{14}C or D, in HOAc, HCOOH and CF$_3$COOH buffered with the presence of the conjugate base of the solvent.[24,25] On the other hand, the scrambling observed in the unbuffered acetolysis of 13a-1-^{14}C or 13b-1-^{14}C was 32-36%.[26,27]

PhCH$_2$CH$_2$X

13

a   b

X = OTs, OTf

CH$_2$—CH$_2$

14

These comparisons point to differences in the solvolytic mechanisms. In solvolyses of 2-arylethyl systems, besides the occurrence of returns to covalent starting material from scrambled ionic species, the net scramblings observed in the products resulted from a combination of a direct displacement, the k_s process, and reaction via the aryl-bridged ion such as the ethylenebenzenium ion 14, the k_Δ process.[26-28] In vinylic systems, a direct S$_N$2 displacement is energetically unfavorable,[29] and where an inversion component has been observed, the intervention of an ion-pair with shielding effects rather than a backside S$_N$2 displacement has been suggested as the mechanism.[30-31]

In the solvolytic studies with 10e-2-^{14}C, while the possibility of returns was not investigated specifically, such processes likely could take place since returns have been observed in studies with triphenyl-vinyl bromide (10g) (Section II.B.3). In this system, however, there should be no S$_N$2 displacement and phenyl-bridging would also be unlikely.[19] The overall extent of scrambling would be largely dependent on the competition between the 1,2-phenyl shift in the triphenylvinyl cation or ion-pair, k_r, and its capture to give product via reaction with solvent, k_{SOH}.

It is also of interest to note that the scramblings found in the acetolysis and formolysis of 10e-2-^{14}C were not very different (6.7 and 7.7%), although there is a large difference in the ionizing power of these two solvents.[32] It was suggested[20] that the extents of scrambling may be closely related to the nucleo-

philicity of the solvents, the nucleophilicity of HOAc and HCOOH being
quite similar and that of CF_3COOH much lower.[33,34] Apparently, the lower
the nucleophilicity of the solvent, the longer will be the lifetime of
the ionic species and the greater will be the extent of scrambling.

B. Studies with triphenyl[2-^{14}C]vinyl or triphenyl[2-^{13}C]vinyl
     bromide (10g-2-^{14}C or 10g-2-^{13}C)

  1. Reaction of 10g-2-^{14}C or 10g-2-^{13}C with HOAc-AgOAc

The reaction of 10g-2-^{14}C or 10g-2-^{13}C with HOAc in the presence
of 1.1 equiv. of AgOAc gave an acetate product, 10b, with 6-7%
scrambling of the label from C-2 to C-1.[35] This work was the first
application of ^{13}C as label, coupled with ^{13}C n.m.r. analysis, in
determining the degenerate rearrangement in a triarylvinyl cationic
system. The agreement between the results from using ^{14}C and ^{13}C as
label confirmed the utility of the quantitative ^{13}C n.m.r. analysis.
A more detailed illustration of this method will be given in
Section III.A.1.

Coincidentally, the 6-7% scrambling arising from 1,2-phenyl shifts
in the reaction of bromides 10g-2-^{14}C or 10g-2-^{13}C with HOAc-AgOAc
was essentially the same as that observed in the acetolysis of the
triflate 10e-2-^{14}C.[20] These results thus appear to indicate that
the acetolysis of the bromide or the triflate gave the same extent
of rearrangement, suggesting that the leaving group has no effect,
and this would not be expected if ion-pairs were involved. A study
on the solvolysis of 10g-2-^{14}C in pure HOAc without any added AgOAc
is impractical since the reaction is too slow and prolonged heating
would cause extensive decomposition. When 10g-2-^{14}C was solvolyzed
in aqueous HOAc, as discussed in the following Section, II.B.2, a
gradual increase in the extent of scrambling from 9.6 to 18.4% was
observed when the amount of HOAc was increased from 50 to 90%. If
these data were extrapolated, one would obtain about 20% scrambling
for the solvolysis of 10g-2-^{14}C in 100% HOAc. This value is
certainly higher than the amount of scrambling observed in the
acetolysis of the triflate 10e-2-^{14}C (6.7%), thus indicating that
the leaving group does affect the extent of rearrangement.

For the reaction of 10g-2-^{14}C or 10g-2-^{13}C in HOAc-AgOAc,
Professor Z. Rappoport[36] has suggested to us that the species
involved may be the quadrapole 15, with the acetate anion intimately
associated with the cationoid moiety of the polarized substrate. A
similar quadrapolar type of interaction has been proposed by
Kernaghan and Hoffmann[37] in accounting for their observation of a

R$^+$ OAc$^-$

Br$^-$ Ag$^+$

$\underline{15}$ R = triphenylvinyl

net retention of the cis- or trans-configuration in the reaction of
cis- or trans-1-phenylpropen-1-yl bromide with silver trifluoroacetate
in isopentane (e.g. eq. 4). The involvement of a quadrapole $\underline{15}$

$$\underset{H}{\overset{CH_3}{>}}C=C\underset{Br}{\overset{Ph}{<}} \xrightarrow[\text{isopentane}]{CF_3COOAg} \underset{H}{\overset{CH_3}{>}}C=C\underset{OOCCF_3}{\overset{Ph}{<}} + \underset{CH_3}{\overset{H}{>}}C=C\underset{OOCCF_3}{\overset{Ph}{<}} \qquad (4)$$

relative ratio: 1.3   1.0

conceivably could give rise to an acetate product, $\underline{10b}$, with a lesser
extent of scrambling (6-7%) than the amount that would have been
observed from solvolysis of $\underline{10g}$-2-^{14}C in pure HOAc (about 20%).

2. Solvolysis of $\underline{10g}$-2-^{14}C in aqueous HOAc

Since solvolysis of bromide $\underline{10g}$ in glacial HOAc is too slow, a study
was made on the extents of isotopic scrambling during the solvolysis
of $\underline{10g}$-2-^{14}C in aqueous HOAc.[38] The solvent systems used varied from
50% HOAc-50% H_2O to 90% HOAc-10% H_2O (by volume). The reactions were
carried out in sealed tubes at 150 $\pm$ 2° with reaction times ranging
from 1-8 days. First of all, in those experiments from which some
unconsumed reactant was recovered and analyzed, there was essentially
no scrambling in the recovered $\underline{10g}$-2-^{14}C, indicating either no ion-
pair return or that the return process could not compete with rearrangement. Such a finding for solvolytic reactions in a water-containing
solvent is not surprising since it is known that, in highly nucleophilic or highly ionizing solvents, the return process competes
poorly with the product forming reaction.[21,22] Kinetic studies with
$\underline{10g}$ by Rappoport and Houminer[19] have also shown that in an aqueous
solvent, namely, 60% EtOH - 40% H_2O, the specific rate constant
remained unchanged throughout a given run, indicating no common ion
rate depression by the liberated bromide ion.

The extent of rearrangement in the reaction product, found to be
solely the unrearranged and rearranged ketones $\underline{16}$-2-^{14}C and $\underline{16}$-1-^{14}C,$\ne$

$\ne$Recent control experiments showed that any acetate product, $\underline{10b}$, that
may have been formed would have been hydrolyzed to the ketone, $\underline{16}$,
under the experimental conditions.[39]

varied depending on the solvent composition. For a given mixture of

$$Ph_2{}^{14}CH-\overset{\overset{\displaystyle O}{\|}}{C}-Ph$$

$$\underline{16}\text{-}2\text{-}{}^{14}C$$

$$Ph-{}^{14}\overset{\overset{\displaystyle O}{\|}}{C}-CHPh_2$$

$$\underline{16}\text{-}1\text{-}{}^{14}C$$

HOAc and H_2O, however, the amount of scrambling remained essentially
constant, although the reaction time and hence the extent of reaction
was changed. In 70% HOAc, the presence of about 3 equiv. of added
NaOAc or NaBr also did not influence the degree of rearrangement.   The
data are summarized in Table 1.

TABLE 1
Solvolysis of Triphenyl[2-^{14}C]vinyl Bromide ($\underline{10g}$-2-^{14}C) in HOAc-H_2O
at 150 ± 2°.[38]

Solvent (% HOAc)	% Scrambling in product[a]	k_{SOH}/k_r
50	9.6 ± 0.3	8.3
60	12.7 ± 0.6	5.9
70	15.3 ± 0.8[b]	4.5
80	16.7 ± 0.2	4.0
90	18.4 ± 0.1	3.4

[a]Mean value ± maximum deviation from the mean; products obtained
at reaction times ranging from 1-8 days, and the number of
individual experiments in each solvent ranged from 2-7.

[b]Include data from experiments carried out with the presence of
about 3 equiv. of NaOAc or NaBr.

The results from the solvolysis of $\underline{10g}$-2-^{14}C in HOAc-H_2O can be
explained by the generalized mechanism given in Scheme 1. With

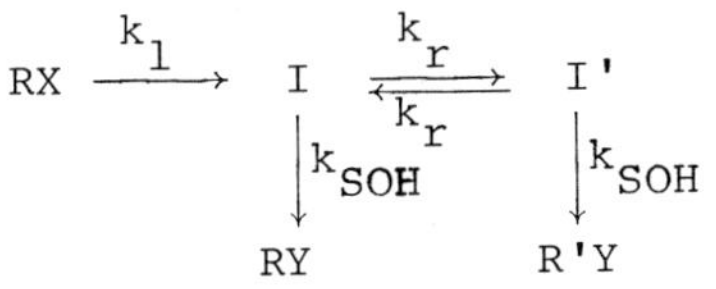

$$RX \xrightarrow{k_1} I \underset{k_r}{\overset{k_r}{\rightleftarrows}} I'$$

with k_{SOH} down from I to RY, and k_{SOH} down from I' to $R'Y$

SCHEME 1

triphenyl[2-^{14}]vinyl bromide ($\underline{10g}$-2-^{14}C) as the substrate, RX, the
unrearranged and rearranged products, RY and R'Y, would be $\underline{16}$-2-^{14}C

and $\underline{16}$-1-^{14}C, respectively. The ionic intermediates I and I',
interconverted by 1,2-phenyl shifts, may be ion-pairs since the
presence of 3 equiv. of NaOAc or NaBr did not affect the extent of
scrambling for the reaction in 70% HOAc. A steady state treatment of
Scheme 1 gives rise to eq. 5 and integration after complete reaction,
using the method of Collins and Bonner,[40] gives eq. 6. Since the

$$(d[RY]/dt)/(d[R'Y]/dt) \quad = \quad 1 + (k_{SOH}/k_r) \tag{5}$$

$$[RY]/[R'Y] \quad = \quad 1 + (k_{SOH}/k_r) \tag{6}$$

ratio of the rates of formation of the unrearranged and rearranged
products is a constant (eq. 5), the percentage of scrambling should
not change with the extent of reaction, and this was as observed.
Utilizing eq. 6, k_{SOH}/k_r can be calculated and these values are
included in Table 1.

As depicted in Scheme 1, the extent of scrambling in the reaction
product would be dependent on the competition between k_r and k_{SOH}.
The higher the proportion of the more nucleophilic H_2O[33,34] in the
solvent mixture, the more effectively will the product forming
process compete with the 1,2-phenyl shift, hence giving rise to a
lesser amount of scrambling. The calculated values of k_{SOH}/k_r given
in Table 1 also reflect this trend.

As has been discussed, the insensitivity of the extent of scrambling
in the triphenylvinyl system towards an added nucleophile, such as the
acetate ion, has been interpreted as suggesting that the 1,2-phenyl
shift may have occurred at the ion-pair stage. However, Rappoport[36]
has pointed out that this conclusion may not necessarily be valid
since the relative concentrations of the solvent and the added lyate
ion should be taken into account. For example, in the solvolysis of
$\underline{10g}$-2-^{14}C in 70% HOAc in the presence of 3 equiv. of NaOAc, the
acetate concentration was about 0.08 M, while for 70% HOAc, $[H_2O]$
and [HOAc] are 16.7 and 12.2 M, respectively. Since the solvent
concentration is much higher, capture of the ionic intermediate by
solvent may be dominant both in the absence and in the presence of
the added OAc$^-$; hence the observed results may not exclude rearrangement via dissociated ions. Further experiments have subsequently
been carried out on the solvolysis of $\underline{10g}$-2-^{14}C in 70% HOAc with up
to 30 equiv. of added NaOAc. The results again showed no appreciable
effect on the extent of scrambling (14.7 ± 0.7%), and analysis of the
data led to the conclusion that ion-pairs were indeed involved in the

triphenylvinyl cationic system.[41]

It is of interest to indicate briefly at this point that solvolyses of tri-p-tolyl[2-^{13}C]vinyl bromide (Section IV.A) in 70% HOAc in the presence of 10 or 30 equiv. of NaOAc did cause a decrease in scrambling when compared with the result observed with no added salt. From the α-substituent effect, the triphenylvinyl cation would be less stable than the tri-p-tolylvinyl cation. It is plausible that while the tri-p-tolylvinyl system may ionize to the dissociated ion, in the triphenylvinyl case, reaction may only proceed to the ion-pair stage, an earlier intermediate than the dissociated ion in Winstein's solvolysis scheme.[42]

3. Solvolysis of 10g-2-^{14}C in 2,2,2-trifluoroethanol (TFE)

The solvolysis of 10g-2-^{14}C in TFE, carried out in the presence of 2,6-lutidine, resulted in isotopic scramblings of the label from C-2 to C-1 in both the reaction product and the recorded reactant.[38] The results are summarized in Table 2. The observation of rearrangements

TABLE 2

Data from Trifluoroethanolysis of 0.12 M Triphenyl[2-^{14}C]vinyl Bromide (10g-2-^{14}C) in the Presence of 0.36 M 2,6-Lutidine at 150 ± 2°C With or Without Added Tetraethylammonium Bromide[38]

Reaction time (days)	% Scrambling from C-2 to C-1		
	Reaction product	Recovered reactant	Mean
1		20.2; 19.9[a]	20.1
2	30.5; 30.4[a]	29.7; 30.3	
		30.2[a]; 29.1[b]	30.0
4	39.9; 39.3[a]	40.3; 40.4[a]	40.0
20	45.0; 44.0[a]		
	43.1[b]		44.1

[a] In the presence of 0.19M Et$_4$NBr.

[b] In the presence of 0.38M Et$_4$NBr.

in the recovered unconsumed reactant is a definitive indication of the occurrence of returns from scrambled ionic intermediates to covalent starting material. This finding is in agreement with the

kinetic observation of Rappoport and Houminer,[19] who reported that
the solvolysis of 10g in TFE/lutidine showed a common ion rate
depression during a given kinetic run or when external bromide ion
was added, indicating the occurrence of return processes.
Incidentally, an earlier observation of an internal return in a vinyl
cationic system was made by Hanack et al.[43] in the solvolysis of
(1-bromo-1-methylmethylene)cyclopropane (17), which gave the
rearranged 1-bromo-2-methylcyclobutene (18) as the main product.

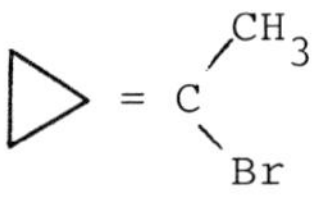

17 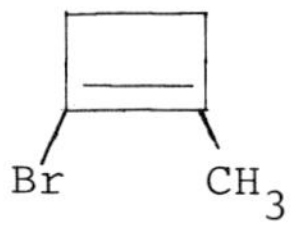   18

   The use of isotopic tracers in observing degenerate rearrangements
in the recovered reactant is an effective way of detecting return
processes, and the method has been used in our laboratory to observe
returns in saturated systems such as in the solvolyses of 2-phenyl
$[1-^{14}C]$ethyl tosylate ($\underline{13a}-1-^{14}C$).[44,45] In the present case with
$\underline{10g}-2-^{14}C$, since added Et_4NBr did not affect the extent of scrambling
in either the product or recovered reactant (Table 2), the involve-
ment of ion-pairs is again indicated. However, there is no experimental evidence to differentiate, according to Winstein's scheme,[42]
whether the ion-pair is an intimate or a solvent-separated ion-pair.
The return observed, therefore, cannot be specified as internal
return or external ion-pair return, but simply as an ion-pair return.
Since common ion rate depression, implicating dissociated ions, has
also been observed in the trifluoroethanolysis of 10g,[19] the ion-pair
presumably could, at least partially, dissociate further to free ions.
   The solvolytic mechanism given in Scheme 1 has to be modified to
include the return process for the trifluoroethanolysis of $\underline{10g}-2-^{14}C$.
Such a mechanism is depicted in Scheme 2, again utilizing generalized

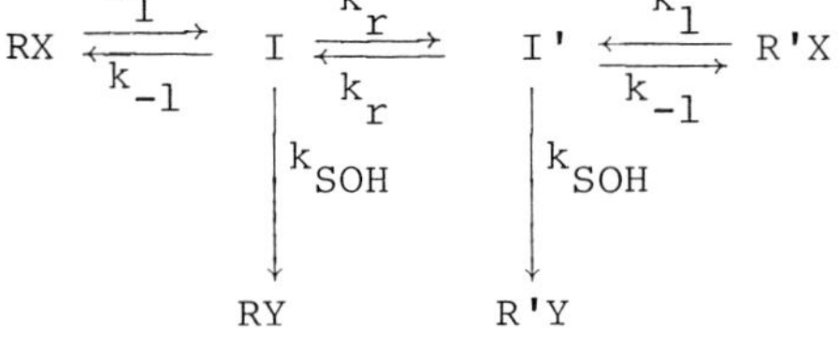

$$RX \underset{k_{-1}}{\overset{k_1}{\rightleftarrows}} \quad I \underset{k_r}{\overset{k_r}{\rightleftarrows}} \quad I' \underset{k_{-1}}{\overset{k_1}{\rightleftarrows}} R'X$$

$$\Big\downarrow k_{SOH} \qquad\qquad \Big\downarrow k_{SOH}$$

$$RY \qquad\qquad R'Y$$

SCHEME 2

symbols rather than the structural formulas. With $\underline{10g}$-2-^{14}C as substrate RX in the trifluoroethanolysis, R'X is the isotopically rearranged $\underline{10g}$-1-^{14}C, the products RY and R'Y are the unrearranged and rearranged triphenyl[2-^{14}C]vinyl and triphenyl[1-^{14}C]vinyl trifluoro-ethyl ethers, and the ionic intermediates I and I' again may be ion-pairs. A steady state treatment for the processes in Scheme 2 gives eq. 7. Since the formation of the unrearranged and rearranged

$$\frac{d[R'Y]/dt}{d[RY]/dt} = \frac{(k_{-1} + k_{SOH})[R'X] + k_r([RX] + [R'X])}{(k_{-1} + k_{SOH})[RX] + k_r([RX] + [R'X])} \tag{7}$$

products would depend on the concentration of the reactant, the extents of scrambling should be different for different reaction times, and this was as observed (Table 2).

It is of interest to note from Table 2 that in the experiments with reaction times of 2 or 4 days, the extent of scrambling in the product and in the recovered reactant were essentially the same. According to eq. 7, if k_r ([RX]+[R'X]) is negligible when compared with $(k_{-1}+k_{SOH})$[R'X] or with $[k_{-1}+k_{SOH})$[RX], then [R'Y]/[RY]=[R'X]/[RX], i.e., the scrambling would be the same in both the product and the recovered reactant. The above interpretation, however, may not hold at very early stages of the reaction since [R'X] arising from the return process would not have been very high. In the 2 or 4 day experiments, the extents of reaction were about 30-60% complete, and apparently [R'X] was built up sufficiently so that k_r([RX] + [R'X]) was negligible when compared to $(k_{-1} + k_{SOH})$[R'X]. The significance of these results is that in the triphenylvinyl system, k_r for the 1,2-phenyl shift is relatively unimportant when compared with k_{-1} and k_{SOH} for the return and product forming processes. Hence once I' is formed, the ratio of [I']/[I] will determine the ratio of their products, [R'Y]/[RY] as well as [R'X]/[RX], because the degree of interconversion between I and I', the k_r process, is comparatively small.

III. THE TRIANISYLVINYL SYSTEM

The first study on the trianisylvinyl cation was reported in 1969 by Rappoport and Gal.[46] Solvolyses, in 80% EtOH, of trianisylvinyl bromide and chloride (19a and 19b) along with 1-anisyl-2,2-diphenyl-vinyl bromide (20) were carried out. The kinetic data led to the conclusion that these reactions proceeded via the intermediate triarylvinyl cation and that there was no anchimeric assistance due

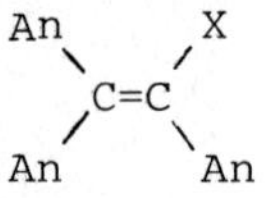

$$\underset{\text{An}}{\overset{\text{An}}{\diagdown}} C = C \overset{X}{\diagup} \qquad X = \text{Br} \quad \text{Cl} \quad \text{OAc}$$

X = Br Cl OAc
 a   b   c

CF₃COO   NNNHPh
 d   e

19

$$\underset{\text{Ph}}{\overset{\text{Ph}}{\diagdown}} C = C \overset{Br}{\underset{\text{An}}{\diagup}}$$

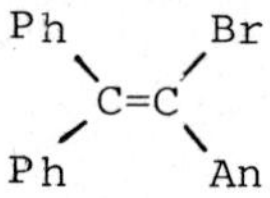

20

to β-aryl participation since 19a and 20 solvolyzed with similar
rates. The reactivity of 19a, 19b and 20 in HOAc containing AgOAc
was found to be much enhanced and vinylic acetates were obtained as
product in high yield. Additional solvolytic studies with trianisylvinyl substrates by Rappoport's group[47-49] all provided further
kinetic results in support of the intermediacy of the trianisylvinyl
cation.

   It may also be worthwhile to note the interesting finding of Sonoda
et al.[50] that if triarylvinyl cations with a β-o-methoxyphenyl,
β-o-methylthiophenyl or β-o-N,N-dimethylaminophenyl substituent were
generated, cyclized products would result. For example, reaction of
Z- or E-2-o-anisyl-1,2-di-p-anisylvinyl bromide (21) in basic (3.3
equiv. NaOH) 80% EtOH gave 2,3-bis-p-anisylbenzofuran (22)
quantitatively. Apparently, the reaction proceeded via triarylvinyl

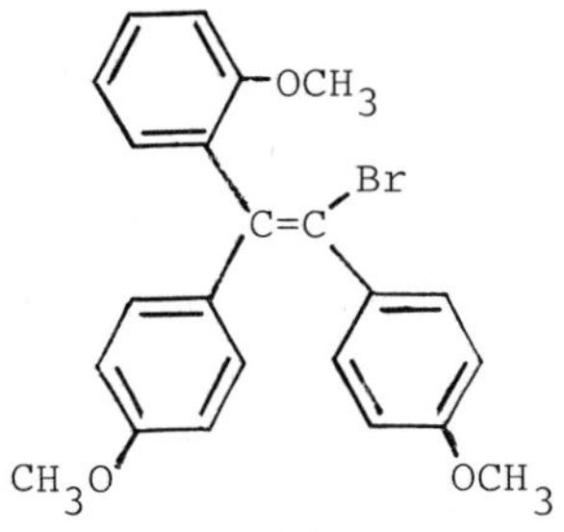

21

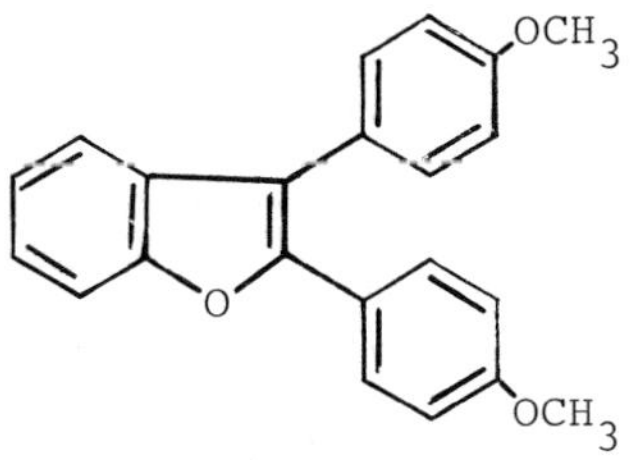

22

cation 23 which cyclized to give oxonium ion 24 followed by
elimination of CH₃ to give 22.

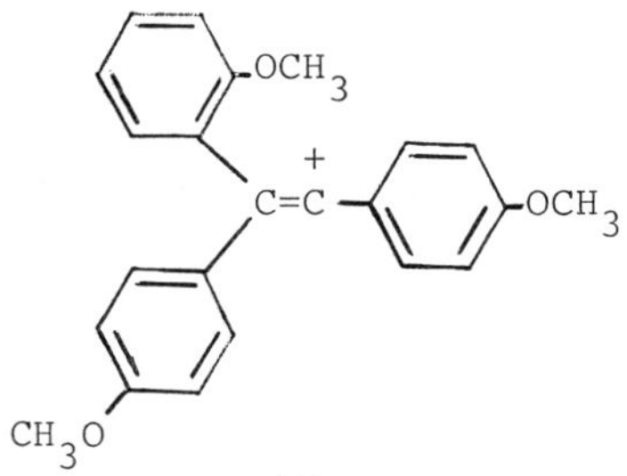

23

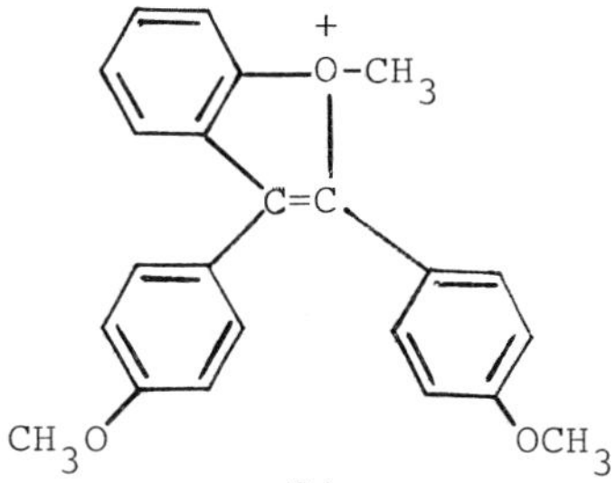

24

A. Reaction of trianisyl[2-^{13}C]vinyl bromide ($\underline{19a}$-2-^{13}C) with HOAc-AgOAc or with CF_3COOH-CF_3COOAg

The first example of a degenerate rearrangement arising from 1,2-anisyl shifts in the trianisylvinyl cation was reported from this laboratory in 1975.[51] From reactions of $\underline{19a}$-2-^{13}C with HOAc in the presence of AgOAc, or with CF_3COOH in the presence of CF_3COOAg, the scramblings of the label from C-2 to C-1 found in the products, trianisyl[1,2-^{13}C]vinyl acetate ($\underline{19c}$-1,2-^{13}C) and trianisyl[1,2-^{13}C]-vinyl trifluoroacetate ($\underline{19d}$-1,2-^{13}C), were 20% and 50%, respectively.

1. Analysis by ^{13}C n.m.r.

It is of interest to include at this point a brief discussion on the analysis of the scrambling of the ^{13}C-label by ^{13}C n.m.r. The methodology and mechanistic applications of ^{13}C n.m.r. have been reviewed by Hinton, Oka and Fry in Volume 3 of this series.[52] For the measurement of degenerate rearrangements in triarylvinyl systems, we utilized a method based on the relative intensities of the ^{13}C n.m.r. absorptions with respect to a suitable internal standard.[53] In studies with $\underline{19a}$-2-^{13}C, the CH_3O absorption containing ^{13}C in its natural abundance was used as such an internal standard. Before being analyzed, the quaternary vinylic carbons have to be converted to saturated ones so that their absorptions will show nuclear Overhauser enhancement. Thus a reaction product, such as trianisyl[1,2-^{13}C]vinyl acetate ($\underline{19c}$-1,2-^{13}C), was reduced to 1,2,2-trianisyl[1,2-^{13}C]ethanol ($\underline{25}$-1,2-^{13}C) before it was analyzed. The ^{13}C absorptions of interest

```
An         OH
  \       /
   CH-CH
  /       \
An         An

   25
```

are those of C-1, C-2 and CH_3O, the internal standard, with their intensities designated as I_1, I_2 and I_S, respectively.

As an illustration of the actual data, for the reaction with HOAc-AgOAc[51] using a sample of $\underline{19a}$-2-^{13}C prepared from [^{13}C]$BaCO_3$ with 45% enrichment, the intensity ratios derived from an unenriched sample of $\underline{25}$ gave I_1^0/I_S and I_2^0/I_S as 0.586 and 0.982, respectively. From the $\underline{25}$-1,2-^{13}C derived from the reaction product, the values of I_1/I_S and I_2/I_S were 5.43 and 33.7, respectively. For C-1, we have

$$I_1^0/I_S = 0.586$$

$I_1/I_S = (I_1^0 + I_1^*)/I_S = 5.43.$

$(I_1^0 + I_1^*)/I_1^0 = 5.43/0.586 = 9.30.$

Since the natural abundance of ^{13}C is 1.1%, $(1.1 + I_1^*)/1.1 = 9.30.$

Hence the enrichment at C-1, $I_1^* = 9.1\%.$

Similar calculations gave $I_2^* = 36.6\%.$

Total ^{13}C enrichment = 36.6 + 9.1 = 45.7%.

Scrambling from C-2 to C-1 = (9.1/45.7) x 100 = 20%.

It has been indicated earlier (Section II, B.1) that triphenylvinyl bromide labelled at C-2 with ^{14}C or ^{13}C, on reaction with HOAc-AgOAc, gave essentially the same extent of scrambling. Similarly, comparisons with trianisylvinyl bromides labelled at C-2 with ^{14}C or ^{13}C also gave scrambling results that were in agreement with each other.[54]

B. <u>Decomposition of trianisyl[2-^{14}C]vinylphenyltriazene (19e-2-^{14}C) in HOAc</u>

It has been pointed out in Section II.A that the decomposition of triphenyl[2-^{14}C]vinylphenyltriazene (10a-2-^{14}C) in a number of acids did not cause any isotopic scrambling in the reaction products.[23] From the same study,[23] it was found that the decomposition of trianisyl[2-^{14}C]vinylphenyltriazene (19e-2-^{14}C) in HOAc gave about 38% scrambling of the label from C-2 to C-1 in the resulting trianisyl[1,2-^{14}C]vinyl acetate (19c-1,2-^{14}C). Interestingly, the presence of 1.7 equiv. of added NaOAc (0.24 M) in the reaction mixture decreased the extent of scrambling to about 17%. These findings indicate that the 1,2-anisyl shift took place in the dissociated trianisylvinyl cation, in contrast to the triphenylvinyl cationic system in which the 1,2-phenyl shift apparently occurred in the ion-pair stage (Section II).

Using the data from the decomposition of 19e-2-^{14}C in the absence or presence of added NaOAc, Rappoport et al.[55] have calculated a k_{OAc^-}/k_{HOAc} ratio for the capture of the trianisylvinyl cation as 19M^{-1}. In the calculation,[36] applying eq. 6 to the scrambling of 38 and 17% observed in the absence and in the presence of 0.24 M NaOAc, the following values can be obtained.

$$k_{HOAc}/k_r = 0.63 \tag{8}$$

$$(0.24\ k_{OAc^-} + k_{HOAc})/k_r = 3.9 \tag{9}$$

Division of eq. 9 by eq. 8 gives a k_{OAc^-}/k_{HOAc} ratio of 22 M^{-1}. The reported somewhat lower value of 19 M^{-1} apparently was due to an error in estimating the concentration of the NaOAc added in the original experiment.[23] It is to be noted that since the concentration of HOAc in pure HOAc is 17.5 M, the dimensionless ratio of k_{OAc}^-/k_{HOAc} should be 17.5 x 22 = 3.8 x 10^2. As this ratio is derived from data obtained in the decomposition of the triazene, the question arises as to whether the same ratio can be applied to the capture of the trianisylvinyl cation generated from solvolyses. Interestingly, a rough estimate using data from the acetolysis of 19a-2-^{14}C does give a k_{OAc}^-/k_{HOAc} value of similar magnitude (Section III.C.2).

C. Solvolytic studies with labeled trianisylvinyl bromide

1. Studies with 1,2-dianisyl-2-p-($^2H_3$)methoxyphenylvinyl bromide (19a-2-methoxy-d_3)

Rappoport and coworkers[56] have carried out extensive degenerate rearrangement studies using deuterium to label the methoxy group of one of the β-anisyl substituents. The substrate, 19a-2-methoxy-d_3, was treated under a variety of solvolytic conditions and the scrambling of the labelled anisyl group was analyzed by ^{1}H n.m.r. and by mass spectrometry. Because of various complications, it was concluded that the n.m.r. method gave the more reliable results. The data obtained are summarized in Table 3.

The various solvents used in this study differ markedly in the combination of the two properties relevant to the rearrangement, namely, their nucleophilicity, N, and their dissociation power which may be measured by the dielectric constant ε. An increase in ε will increase the stability and lifetime of the cation and a decrease in N will decrease k_{SOH}, allowing the rearrangement process, k_r, to compete more effectively with product formation. 2,2,2-Trifluoroethanol (TFE) has high dissociation power (ε = 26.67[57]) and low nucleophilicity (N = -2.78[34]) and is judged to be the best solvent in promoting rearrangement, as can be seen from the data in Table 3. Although CF$_3$COOH also gave 100% rearrangement, it is judged to be an unsuitable solvent for solvolytic studies with triarylvinyl substrates because of the possibility that, in this medium, scrambling may occur after the formation of the initial product. Moreover, there may also be addition-elimination processes in which an initial protonation to give sp^3-hybridized ions which migh subsequently rearrange (see Section IV.A).

The suppression of the rearrangement observed upon addition of the common bromide ion was taken as clear and strong evidence that the reaction of 19a-2-methoxy-d_3 proceeded via the dissociated trianisyl-

TABLE 3

Degenerate Rearrangements During Solvolyses of 1,2-Dianisyl-2-p-(^{2}H$_3$)methoxyphenylvinyl bromide (19a-2-methoxy-d$_3$)

Solvent	Added base	Added salt	% Rearrangement[a] From n.m.r.	From m.s.
60% EtOH	NaOAc		11.5±2	16.5±1
HOAc	NaOAc		35±2	51±6
HOAc	NaOAc	Bu$_4$NBr	4±1[b]	
1:1 HOAc-HCOOH	NaOAc		40±2	73±5
CH$_3$COOH	NaOOCCF$_3$			100
TFE (8 hr.)	2,6-lutidine		100±3	
TFE (14 min.)	2,6-lutidine		>95; >80±15[b]	
TFE (14 min.)	2,6-lutidine	Bu$_4$NBr (0.8M)	56±3[b]	
TFE (14 min.)	2,6-lutidine	Bu$_4$NBr (3.6M)	0±3[b]	
(CH$_3$)$_3$CCOOH			0±3[b,c]	
CH$_3$CN (50 hr.)			15±1[b,c]	
CH$_3$CN (160 hr.)			43±1[b,c]	

[a] Complete 100% rearrangement corresponds to the formation of 33.3% of the scrambled 2,2-dianisyl-1-p-(^{2}H$_3$)methoxyphenylvinyl product; on this basis, the % rearrangement corresponds to 2 x % scrambling obtained with trianisylvinyl substrates labeled at C-2 with ^{13}C or ^{14}C.
[b] In the recovered vinyl bromide.
[c] No solvolytic substitution product.

vinyl cation in both TFE and HOAc. Since some rearrangement was observed in the recovered, unconsumed bromide in the acetolysis and trifluoroethanolysis, the generalized mechanism outlined in Scheme 2 (page 12) should apply, with trianisylvinyl bromide as the substrate RX and the unrearranged and rearranged intermediates I and I' as dissociated trianisylvinyl cations. However, since one of the anisyl groups is labelled, there is a statistical factor and Scheme 2 may be modified to give Scheme 3 specifically for the case of 19a-2-methoxy-d$_3$.

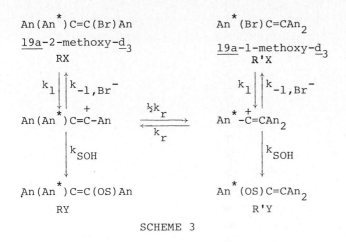

SCHEME 3

When the reaction is complete, there is no remaining bromide. The infinity product distribution thus is not affected by the Br^- capture and the k_{-1} process can be neglected. Integration by the method of Collins and Bonner[40] would give eq. 10, a modified version of eq. 6 (page 10). Using the "more reliable" value from n.m.r. of $35 \pm 2\%$

$$[RY]/[R'Y] = 2[1 + (k_{SOH}/k_{r(An)})] \qquad (10)$$

rearrangement for the acetolysis in the presence of NaOAc of 19a-2-methoxy-$\underline{d}_3$, Rappoport et al.[56] calculated using eq. 10 a value of 2.8 ± 0.25 for $k_{SOH}/k_{r(An)}$. From the 20% scrambling (40% rearrangement) obtained by Oka and Lee[51] in the reaction of 19a-2-^{13}C and applying eq. 6, $k_{SOH}/k_{r(An)}$ was found to be 3.0 ± 0.25. It was concluded that this agreement lends further support to the mechanism involving dissociated ions, as given in Scheme 3.

With regard to the above comparison, the following considerations, however, should be noted. In the two sets of experiments being compared, one involved reaction with HOAc containing about 1.1 mol equiv. of AgOAc and the other with HOAc containing 2 mol equiv. of NaOAc. If k_{SOH} is the sum of $(k_{HOAc} + [OAc^-]k_{OAc}-)$ (Section III.B) and since the acetate ion concentrations are different, the two experiments should not be compared directly. Moreover, any effect of the Ag^+ ion has not been taken into account. The apparent agreement in the $k_{SOH}/k_{r(An)}$ values calculated from the two sets of data may, therefore, be just coincidental. Interestingly, in the solvolysis of 19a-2-^{14}C in 9:1 HOAc-Ac$_2$O containing 2 mol equiv. of NaOAc, to be discussed in the following section (III.C.2), after the reaction is about 90% complete, about 24% scrambling (48% rearrangement) was observed. Extrapolation to 100% reaction gave roughly 27% scrambling or 54% rearrangement, which is in better agreement with the 51 ± 6% rearrangement obtained by Rappoport et al.[56] using what was believed to be the less reliable method of mass spectral analysis.

Rappoport and coworkers[9,58] have observed unusually high degrees of common ion rate depression in solvolyses of triarylvinyl bromides, and it was proposed that this high selectivity of the triarylvinyl cation, as measured by $\alpha(k_{-1}/k_{SOH}$ or $k_{-1}/k_{SO}-)$, is not only due to the stabilization of the cation by the α-aryl substituent. The data indicated that the more bulky the β-substituents in the α-anisyl substituted vinyl cation, the higher is its selectivity factor α. A shielding of the cationic orbital by the bulky substituents in a "static" vinyl cation was suggested as chiefly responsible for this high selectivity. With the p-orbital of the triarylvinyl cation sterically shielded from approach by nucleophiles, the lifetime of the cation as well as its selectivity will increase in parallel with the bulk of the β-substituents, since the more polarizable bromide ion will capture the triarylvinyl cation more effectively than the solvent or lyate ion.[9,58]

Based on some of the data in Table 3, Rappoport et al.[56] have discussed and rejected an alternative explanation for the high selectivity of ions such as the trianisylvinyl cation. In the trifluoroethanolysis of 19a-2-methoxy-d_3, for example, the rearrangement in the reaction product is complete and the three aryl groups in the substrate are all anisyl groups. Rapid back and forth shiftings between C-1 and C-2 by these anisyl groups together with appropriate rotations of the three groups could then take place on both sides of the vinylic double bond. This in effect will result in a rapid "spherical movement" of the three aryl groups which could drive off solvent and bromide ions from the vicinity of C-1 and C-2, causing a "3-dimensional windshield wiper effect", which presumably might increase the lifetime and hence the selectivity of the cation.[36,56]

If the windshield wiper effect described above were to contribute to the high selectivity, an additional condition must be met; namely, the rearrangement must also be faster than capture of the trianisylvinyl cation by bromide ion, i.e. $k_{r(An)} >> k_{-1}[Br^-]$, even with added external Br^-. The data in Table 3 for trifluoroethanolysis show that when Br^- is added, the trianisylvinyl cation is captured preferentially by Br^- rather than by solvent. At the high Br^-

concentration of 3.6 M, the recovered vinyl bromide was not rearranged. Hence $k_{-1}[Br^-] > k_{r(An)}$, and it follows that the selectivity of the trianisylvinyl cation is not due to a windshield wiper effect.

2. Studies with trianisyl[2-^{14}C]vinyl bromide(19a-2-^{14}C)

Lee et al.[59] have investigated the solvolysis of 19a-2-^{14}C in HOAc-Ac$_2$O (9:1 by volume) in the presence of various amounts of added NaOAc or NaOAc and LiBr. The extents of isotopic scrambling in both the reaction product and recovered reactant at different stages of reaction were determined and the results are summarized in Table 4. These results extend the data obtained by Rappoport

TABLE 4

Scramblings During Acetolysis of 15 mM 19a-2-^{14}C in the Presence of Added NaOAc or NaOAc and LiBr at 120 ± 2°

[NaOAc] mM	[LiBr] mM	Reaction time, hr.[a]	Scrambling from C-2 to C-1, %	
			Reaction product	Recovered reactant
30		5 (25)	9.6	0.7
30		10 (50)	10.1; 10.3	4.9; 3.8
30		70 (75)	17.4; 17.6	20.3; 20.1
30		100 (90)	23.7; 24.1	28.8; 29.9
300		3 (50)	7.7; 7.7	0.3; 0.5
300		30 (95)	10.7; 10.0	5.9; 6.4
30	30	21 (25)	6.7	0.5
30	30	64 (50)	9.1; 8.5	3.9; 4.0
30	30	135 (75)	15.5; 15.0	18.5; 17.6
30	30	384 (85)	38.2; 37.9	49.6; 50.9
30	60	77 (50)	7.8; 7.9	3.7; 3.4
30	60	384 (80)	32.0; 30.6	45.0; 46.5
30	100	135 (50)	6.9; 6.8	2.8; 3.0
30	100	432 (70)	19.8; 22.9	40.2; 38.5
30	300	96[b] (20)	4.4	1.1

[a]The figures in parentheses give the approximate % reaction.
[b]The reaction mixture blackens on further heating.

et al.[56] for acetolysis and support the conclusion that the dissociated trianisylvinyl cation is involved. The mechanism as outlined in Scheme 2 and eq. 7 (page 13) derived for this scheme should apply.

From eq. 7, it is expected that the amount of scrambling in the reaction product would increase with increasing extent of reaction, and this was as observed under each given set of experimental conditions (Table 4). It may be noted that a 10-fold increase in the amount of added NaOAc significantly decreased the extent of scrambling, which, aside from supporting a role for dissociated ions, also provided another estimate for the ratio of $k_{OAc}-/k_{HOAc}$. If the reactions in the presence of 30 and 300 mM NaOAc were extrapolated to completion, the scramblings would correspond to roughly 27% and 11%, respectively. As discussed in Section III.C.1, at complete reaction, the k_{-1} process can be neglected and the integrated eq. 6 can apply. Using eq. 6 and the procedure described in Section III.B, eqs. 11 and 12 can be obtained. From eqs. 11 and

$$(k_{HOAc} + 0.03\ k_{OAc}-)/k_r = 1.7 \tag{11}$$

$$(k_{HOAc} + 0.30\ k_{OAc}-)/k_r = 7.1 \tag{12}$$

12, $k_{OAc}-/k_{HOAc} = 18\ M^{-1}$, giving a dimensionless ratio of $k_{OAc}-/k_{HOAc}$ as 18 x 17.5 = 3.2 x 10^2. This value, although based on scramblings roughly estimated by extrapolation, is of the same order of magnitude as that derived from data obtained in the decomposition of the triazene $\underline{19e}$-2-^{14}C (page 16).

The observed scramblings in the recovered reactant again constituted definitive evidence for the occurrence of returns. The implication of dissociated ions, both by the decrease in scrambling with added NaOAc or LiBr and by the earlier finding of extensive common ion rate depression,[47,48] indicated that the return process may be classified as external ion return. It may also be noted from Table 4 that in the latter stages of reaction, the scrambling in the recovered reactant could be higher than the scrambling in the corresponding reaction product. This finding may be attributed to the occurrence of many cycles of ionization, 1,2-shift and return, which eventually could lead to the accumulation of a greater amount of rearranged reactant in the unconsumed substrate.

It is also of interest to contrast the results in Table 4 with the data obtained from the trifluoroethanolysis of triphenyl[2-^{14}C]-vinyl bromide ($\underline{10g}$-2-^{14}C). In both of these cases, scramblings were observed in the product and in the recovered reactant. In the trifluoroethanolysis of 10g-2-^{14}C, after about 30% and 60% reaction, the scramblings in the product and in the recovered reactant were essentially the same (Table 2, page 11). As was discussed in Section II.B.3, this finding suggested that in the triphenylvinyl system, $k_{r(Ph)}$ for the 1,2-phenyl shift is relatively unimportant when compared with k_{-1} and k_{SOH}. For the trianisylvinyl system, however, $k_{r(An)}$ is not unimportant when compared with k_{-1} and k_{SOH}, and in accordance with eq. 7, the scramblings in the product and in the recovered reactant for a given experiment are not the same (Table 4). This interpretation is, of course, quite reasonable since the migratory aptitude for the 1,2-shift should be much larger for the anisyl than the phenyl group.

IV. THE TRI-p-TOLYLVINYL SYSTEM

A. Studies with tri-p-tolyl[2-^{13}C]vinyl bromide (26a-2-^{13}C)

The reaction of 26a-2-^{13}C with HOAc containing 1.1 equiv. of AgOAc gave a product, tri-p-tolyl[1,2-^{13}C]vinyl acetate (26b-1,2-^{13}C), with 13-14% scrambling of the label from C-2 to C-1.[60] This value,

```
Tol      X       X = Br  OAc  OOCCF3                      Ar
  \     /                                                  ‚ +
   C=C                      a     b      c             Ar-C = C-Ar
  /     \
Tol      Tol

   26                                                      27
```

as expected, is intermediate between the approximately 7% and 20% scramblings, respectively, found for similar reactions with triphenyl[2-^{13}C]vinyl and trianisyl[2-^{13}C]vinyl bromides (10g-2-^{13}C and 19a-2-^{13}C, Sections II.B.1 and III.A).

It is well known that the migratory aptitudes of aryl groups follow the order An > Tol > Ph. From the classical work of Bachmann[61] with symmetrical aromatic pinacols, the migratory aptitudes for Ph:Tol:An was found to be 1:15.7:500. The actual ratio of % scrambling in the reaction with HOAc-AgOAc for 10g-2-^{13}C, 26a-2-^{13}C and 19a-2-^{13}C is 7:14:20 or about 1:2:3. These values are much smaller than one would expect if migratory aptitudes were the predominant factor. It was pointed out[60] that for the three triarylvinyl bromides being compared, however, not only the migrating aryl groups but also the aryl substituents at the migration origin and migration terminus are different from one system to another.

If 27 were used as a model of the transition state for the 1,2-
aryl shift, in going from the triarylvinyl cation to 27, the higher
the stability (lower energy) of the initial vinyl cation, the greater
may be the activation energy required to reach transition state 27,
and this effect would counterbalance the order of migratory aptitudes
of An > Tol > Ph. Thus in order to get a better evaluation of
migratory aptitudes, comparisons of data from systems with different
migrating groups but with the same non-migrating substituents should
be used. An example of such a comparison using data from the tri-p-
tolylvinyl and 2-phenyl-1,2-di-p-tolylvinyl systems will be described
in Section VII.A.

It may also be pointed out that another factor which could also
counterbalance the greater migratory aptitudes of the tolyl and
anisyl groups is the effect of the α-aryl substituent on the
electrophilic character of the migration terminus. In saturated
carbocations generated from deaminations, the An:Ph migratory
ratios have been found to be quite low (about 1.2 to 2.0[62-64]).
Moreover, Bonner and Putkey[65] have actually observed a lower extent
of 1,2-anisyl shift (17%) in the deamination of 1,2,2-trianisyl[1-^{14}C]-
ethylamine than the 1,2-phenyl shift (26-28%) in the deamination of
1,2,2-triphenyl[1-^{14}C]ethylamine. A diminished positive character
due to the delocalization of the positive charge at the migration
terminus has been suggested as playing an important role in
accounting for this reversal in migratory aptitudes.[65-66] Similarly,
in the triarylvinyl cations, the delocalization of the positive
charge by the α-tolyl or α-anisyl substituent would decrease the
electrophilic character of the migration terminus, and hence decrease
the tendency for the tolyl and anisyl migrations.

The reaction of 26a-2-^{13}C with CF$_3$COOH in the presence of 1.1 equiv.
of CF$_3$COOAg gave 47% scrambling of the label from C-2 to C-1 in the
reaction product, 26c-1,2-^{13}C.[60] A similar reaction with 19a-2-^{13}C
has been found to give a complete scrambling of 50% (Section III.A).
As discussed earlier (Section II.A), the extremely low nucleophilicity
of CF$_3$COOH would allow the 1,2-aryl shift in the vinyl cation to
compete favorably with the product forming reaction with solvent,
which is in agreement with the extensive scramblings observed in the
trifluoroacetolysis.

In solvolytic studies with vinylic systems under acidic conditions
such as trifluoroacetolysis, there is the possibility of an electrophilic addition followed by elimination to give the same product as
that derived from an S$_N$1 reaction.[67-70] It has been suggested that

a solvent kinetic isotope effect should be a good diagnostic test
for such an addition-elimination process.[69,70] When 26a was
solvolyzed in HOAc or DOAc at 150°, there was essentially no solvent
isotope effect. On the other hand, trifluoroacetolysis in CF_3COOH
and CF_3COOD at 100°, without any added salt, gave a solvent isotope
effect, $k_{CF_3COOH}/k_{CF_3COOD}$, of 2.6, suggesting a significant role for
electrophilic addition-elimination. Similar reactions carried out in
the presence of 2.5 equiv. of CF_3COONa gave a lower $k_{CF_3COOH}/k_{CF_3COOD}$
value of 1.6. With the presence of 1.1 equiv. of CF_3COOAg, however,
the trifluoroacetolysis was too fast for potentiometric titration and
no solvent isotope effect was determined. It was proposed[60] that,
presumably, the Ag salt could catalyze the formation of the vinyl
cation via the S_N1 process and the contribution from electrophilic
addition-elimination would be negligible, while in the presence of
CF_3COONa, the contribution of the addition-elimination process may
be of lesser importance than in the unbuffered trifluoroacetolysis.

In the unbuffered trifluoroacetolysis of 26a-2-^{13}C at 100°, the
rearrangement in the product was also found to increase with increasing
reaction time. After about 2 and 5 half-lives and at completion, the
scramblings obtained were 43, 47 and 50%, respectively.[60] These
findings thus also indicate the possibility that the initially formed
product in the trifluoroacetolysis might undergo further scrambling
upon being heated in the reaction medium.

As stated in Section II.B.2, in connection with the studies on the
involvement of ion-pairs in the solvolysis of triphenyl[2-^{14}C]vinyl
bromide (10g-2-^{14}C) in 70% HOAc - 30% H_2O, investigations have also
been carried out on the solvolysis of 26a-2-^{13}C in 70% HOAc.[41]
Similar to the results observed with 10g-2-^{14}C, there was no
scrambling in the recovered reactant and the scrambling in the product
did not change with reaction time, indicating that the generalized
mechanism given in Scheme 1 (page 9) and the integrated eq. 6 should
apply to the reaction of 26a-2-^{13}C in 70% HOAc. However, unlike the
behavior of 10g-2-^{14}C, which showed no appreciable change in the
extent of scrambling when the reaction was carried out with added
NaOAc, the reaction of 26a-2-^{13}C in 70% HOAc carried out with the
presence of 10 and 30 equiv. of added NaOAc decreased the extent of
scrambling to a mean of 14.6 and 11.4%, respectively, from a mean
value of 18.6% scrambling observed without any added NaOAc. These
results are interpreted[41] as indicating that in 70% HOAc, 26a-2-^{13}C
ionizes to dissociated ions and 1,2-tolyl shifts take place in the

dissociated tri-p-tolylvinyl cation (I and I' in Scheme 1 as
unrearranged and rearranged and rearranged dissociated ions).

V. THE 2-ANISYL-1,2-DIPHENYLVINYL SYSTEM

Solvolytic studies on cis- and trans-2-anisyl-1,2-diphenylvinyl
bromide (cis- and trans-11a), along with studies on triphenylvinyl
and 2,2-dianisyl-1-phenylvinyl bromides (10g and 12), were reported
by Rappoport and Houminer in 1973.[19]   It was found that reaction of
either cis- or trans-11a in TFE buffered with 2,6-lutidine gave
identical products consisting of about 85% of the structurally
rearranged 1-anisyl-2,2-diphenylvinyl trifluoroethyl ether (28) and
about 15% of a 1:1 mixture of the structurally unrearranged cis- and
trans-2-anisyl-1,2-diphenylvinyl trifluoroethyl ethers (cis- and
trans-11b). On the other hand, solvolysis of either cis- or trans-
11a in the more nucleophilic 60% EtOH gave only about 5% of the
rearranged 1-anisyl-2,2-diphenylethanone (29) and about 95% of the
unrearranged 2-anisyl-1,2-diphenylethanone (30) (Scheme 4).

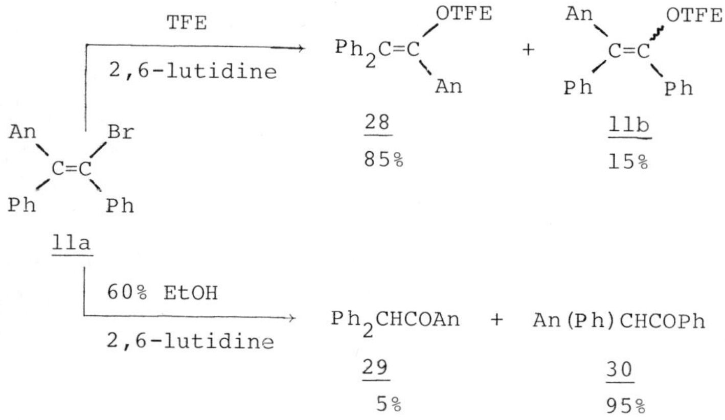

TFE
2,6-lutidine

$Ph_2C=C$ OTFE / An   +   An OTFE / $C=C$ / Ph Ph

28   11b
85%   15%

An Br / $C=C$ / Ph Ph

11a

60% EtOH
2,6-lutidine

$Ph_2CHCOAn$ + $An(Ph)CHCOPh$

29   30
5%   95%

SCHEME 4

It was also found that reaction of trans-11a in HOAc containing
2 equiv. of AgOAc gave exclusively a 55:45 mixture of the unrearranged
acetates, cis- and trans-11c.[71] As stated earlier on page 4, these
results from cis- and trans-11a, together with their kinetic
behaviors and data from 10g and 12, led to the conclusion that
triarylvinyl bromides solvolyze via dissociated vinyl cations
without β-aryl participation.[19]

A. Studies with <u>cis</u>- and <u>trans</u>-2-anisyl-2-(^{2}H$_5$)phenyl-1-phenylvinyl

   bromides (<u>cis</u> and <u>trans</u>-<u>11a</u>-2-Ph-<u>d</u>$_5$)

   Rappoport, Noy and Houminer[72] have carried out degenerate rearrangement studies with <u>11a</u>-2-Ph-<u>d</u>$_5$ using the pentadeuterophenyl group as
a label. From solvolyses, the structurally unrearranged product was
found to show degenerate rearrangements arising from 1,2-anisyl
shifts, thus resulting in the scrambling of the Ph-<u>d</u>$_5$ label from
C-2 to C-1 (eq. 13):

$$\underset{Ph^*}{\overset{An}{\diagdown}}C=C-Ph \; \underset{\longleftarrow}{\longrightarrow} \; Ph^*-C=C\underset{Ph}{\overset{An}{\diagup}} \qquad\qquad (13)$$

   For solvolyses in 60 or 80% EtOH buffered with 2,6-lutidine, the
major product (about 95%) was ketone <u>30</u> with the Ph-<u>d</u>$_5$ label
scrambled over both C-2 and C-1. Similarly, in the reaction with
HOAc-AgOAc, the product was a 55:45 mixture of <u>cis</u>- and <u>trans</u>-<u>11c</u>-1,2-
Ph-<u>d</u>$_5$, again with the labelled Ph-<u>d</u>$_5$ substituent scrambled over C-2
and C-1. Analyses for the extents of scrambling were made by ^{1}H
n.m.r. and by mass spectrometry, with both methods giving results in
reasonably good agreements. The "best data" are summarized in Table 5.

TABLE 5
Extent of Degenerate Rearrangement during the Solvolysis of <u>cis</u>- and
<u>trans</u>-<u>11a</u>-2-Ph-<u>d</u>$_5$

Substrate	Solvent	% Rearrangement[a]
<u>trans</u>-<u>11a</u>-2-Ph-<u>d</u>$_5$	60% EtOH	88.6 ± 1.4
7:3 <u>cis</u>- and <u>trans</u>-		
<u>11a</u>-2-Ph-<u>d</u>$_5$	60% EtOH	90.7 ± 0.5
<u>trans</u>-<u>11a</u>-2-Ph-<u>d</u>$_5$	80% EtOH	73 ± 3
<u>trans</u>-<u>11a</u>-2-Ph-<u>d</u>$_5$	HOAc-AgOAc	92.7 ± 1.8

[a]100% Rearrangement corresponds to a 1:1 distribution of the labeled
Ph-<u>d</u>$_5$ substituent over C-2 and C-1 in the product.

   In the trifluoroethanolysis of <u>trans</u>-<u>11a</u>-2-Ph-<u>d</u>$_5$ buffered with
2,6-lutidine and carried out in the presence of 0.15 M Bu$_4$NBr, the
major product (72%) was the structurally rearranged ether <u>28</u>-2-Ph-<u>d</u>5.

A mixture of <u>cis</u>- and <u>trans-11b</u>-1,2-Ph-d_5 (10%) as well as unconsumed <u>cis</u>- and <u>trans-11a</u>-Ph-d_5 (18%) were detected. Analysis by ^{1}H n.m.r. for degenerate rearrangement in the minor product, <u>cis</u>- and <u>trans-11b</u>-1,2-Ph-d_5, was not feasible, but the mass spectrum suggested that in this product, the extent of degenerate rearrangement was essentially complete.

The finding that solvolysis in 60% EtOH of <u>trans-11a</u>-2-Ph-d_5 as well as a 7:3 mixture of the <u>cis</u>- and <u>trans</u>-isomers gave essentially the same extent of scrambling, arising from 1,2-anisyl shifts (Table 5), was taken as evidence for the formation of a dissociated linear 2-anisyl-1,2-diphenylvinyl cation (<u>31</u>) as intermediate. If the

rearrangement were to proceed via an initial formation of a bridged ion (<u>32</u>), <u>trans-11a</u>-2-Ph-d_5, with its potentially participating anisyl group <u>trans</u> to the leaving Br group, would be expected to give a greater extent of degenerate rearrangement than the <u>cis</u>-isomer, but this was not observed.

If the rearrangement were to occur at the ion-pair stage, similar to the inversion which was formed for vinylic solvolysis via ion-pairs,[30-31] shielding by the leaving group in an ion-pair would decrease the occurrence of 1,2-anisyl shifts on the shielded side. This would also result in more rearrangement from <u>trans-11a</u>-2-Ph-d_5 than the <u>cis</u>-isomer, again contrary to what was observed. Thus the essentially identical extent of degenerate rearrangement from <u>cis</u>- or <u>trans-11a</u>-2-Ph-d_5 and the earlier observation[19,71] of the formation, from either <u>cis</u>- or <u>trans-11a</u>, of a 1:1 mixture of <u>cis</u>- and <u>trans-11b</u> from trifluoroethanolysis, and a similar mixture of <u>cis</u>- and <u>trans-11c</u> from reaction with HOAc-AgOAc, all are consistent with dissociated ion <u>31</u> as reaction intermediate.

In saturated cationic systems, the migratory aptitude of An/Ph in solvolytic type of reaction have been reported to range from about 6-500.[61,73,74] In deaminations, however, the ratio was found to be much lower (1.2 - 2.0[62-64]) or even less than unity.[65,66] Rappoport et al.[72] thus pointed out the possibility that the An/Ph migration ratio may not necessarily be greater than unity in vinyl cationic systems. If a bridged structure such as <u>32</u> were assumed to be a model of the transition state for the 1,2-aryl shift, the gain in the bridging by charge dispersal over the migrating group (An greater than Ph) may be overcome by an opposing effect due to a loss of ground state stabilization by the conjugation of the migrating group and the vinylic double bond (the $\pi(\beta\text{-Ar}) - \pi(C=C)$ conjugation), and this loss would be greater for the An than the Ph group. The An/Ph migration ratio, based on data from the solvolysis of 11a-2-Ph-d_5, has been evaluated[72] by considering the processes in Scheme 5, in which solvolysis in aqueous EtOH is used as illustration.

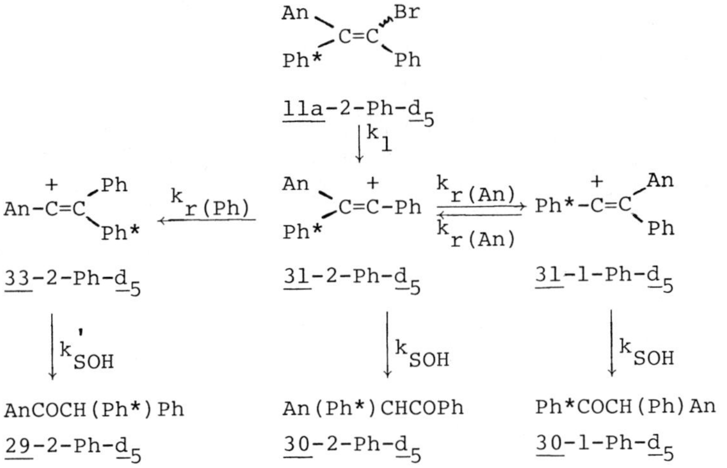

$$\begin{array}{c} \text{An} \\ \diagdown \\ \text{Ph*} \end{array} C=C \begin{array}{c} \text{Br} \\ \diagup \\ \text{Ph} \end{array}$$

11a-2-Ph-d_5

$\downarrow k_1$

$$\overset{+}{\text{An-C=C}}\begin{array}{c}\text{Ph}\\\diagdown\\\text{Ph*}\end{array} \xleftarrow{k_{r(Ph)}} \begin{array}{c}\text{An}\\\diagdown\\\text{Ph*}\end{array}\overset{+}{\text{C=C-Ph}} \underset{\overset{\displaystyle\longrightarrow}{k_{r(An)}}}{\overset{k_{r(An)}}{\rightleftharpoons}} \overset{+}{\text{Ph*-C=C}}\begin{array}{c}\text{An}\\\diagdown\\\text{Ph}\end{array}$$

33-2-Ph-d_5 31-2-Ph-d_5 31-1-Ph-d_5

$\downarrow k'_{SOH}$ $\downarrow k_{SOH}$ $\downarrow k_{SOH}$

AnCOCH(Ph*)Ph   An(Ph*)CHCOPh   Ph*COCH(Ph)An

29-2-Ph-d_5 30-2-Ph-d_5 30-1-Ph-d_5

SCHEME 5

From Scheme 5, using the steady-state treatment and neglecting isotope effects, eqs. 14 and 15 were derived.[72] For solvolysis in

$$[\underline{29}\text{-2-Ph-}\underline{d}_5]/([\underline{30}\text{-2-Ph-}\underline{d}_5] + [\underline{30}\text{-1-Ph-}\underline{d}_5]) = k_{r(Ph)}/k_{SOH} \qquad (14)$$

$$[\underline{30}\text{-2-Ph-}\underline{d}_5]/[\underline{30}\text{-1-Ph-}\underline{d}_5] = 1 + (k_{SOH}/k_{r(An)}) \qquad (15)$$

60% EtOH, since 29 and 30 were formed to the extent of 5% and 95%, respectively,[19] one obtains, from eq. 14, $k_{SOH}/k_{r(Ph)}$ = 19. Using the data in Table 5 (88.6 - 90.7% rearrangement) and eq. 15, $k_{r(An)}/k_{SOH}$ is calculated to be about 4.0. Hence the order of reactivity for $k_{r(An)}:k_{SOH}:k_{r(Ph)}$ in 60% EtOH is 76:19:1. Similarly, in the reaction of 11a with HOAc-AgOAc, although no 29 was detected,[71] it was estimated that the limit for detection by n.m.r. was 5%, and assuming there was up to 5% 29, again $k_{SOH}/k_{r(Ph)}$ = 19. From the degenerate rearrangement of 92.7% (Table 5), $k_{r(An)}/k_{SOH}$ = 6.3, and hence in HOAc-AgOAc, the ratio of $k_{r(An)}:k_{SOH}:k_{r(Ph)}$

is 120:19:1. Thus it was concluded[72] that the migration ratio of An/Ph for the rearrangement processes in cation 31 would be in the range of 76-120.

On checking the integration using the method of Collins[40] for the processes in Scheme 5, an error was found in eq. 14, which should be corrected to give eq. 14a, without the [30-1-Ph-$\underline{d}_5$] term in the denominator on the left side of eq. 14.

$$[\underline{29}\text{-}2\text{-Ph-}\underline{d}_5]/[\underline{30}\text{-}2\text{-Ph-}\underline{d}_5] = k_{r(Ph)}/k_{SOH} \tag{14a}$$

For solvolysis in 60% EtOH, with a degenerate rearrangement of 88.6 - 90.7% (Table 5), the product distribution for 29-2-Ph-$\underline{d}_5$, 30-2-Ph-$\underline{d}_5$ and 30-1-Ph-$\underline{d}_5$ should be about 5, 52 and 43% respectively. From eq. 14a, $k_{SOH}/k_{r(Ph)}$ is about 10, instead of the reported value of 19, and hence the ratio $k_{r(An)}:k_{SOH}:k_{r(Ph)}$ is 40:10:1. For the reaction with HOAc-AgOAc, again assuming formation of 5% 29, with 92.7% degenerate rearrangement (Table 5), the product distribution for 29-2-Ph-$\underline{d}_5$, 30-2-Ph-$\underline{d}_5$ and 30-1-Ph-$\underline{d}_5$ would be about 5, 51 and 44%, respectively. Utilizing eqs. 14a and 15, $k_{SOH}/k_{r(Ph)}$ is again about 10, and $k_{r(An)}:k_{SOH}:k_{r(Ph)}$ is 63:10:1. These corrected values suggest that for solvolyses of 11a, the An/Ph migration ratio is in the range of 40-63, instead of the reported 76-120. Further discussion of migratory aptitudes will be given in Section VII.A.

The above An/Ph migration ratios, in any event, are well above unity. It was concluded[72] that a differential ground state $\pi(\beta\text{-Ar}) - \pi(C=C)$ conjugation between the β-phenyl and β-anisyl groups thus has relatively little influence on the extent of β-aryl rearrangement. This was believed to be not surprising since the two β-aryl groups in an ion such as 31 may interact sterically, causing a twist of both of them from the plane of the vinylic double bond.

In discussing the results from studies with cis- and trans-11a-2-Ph-$\underline{d}_5$, Rappoport et al.[72] also made the comparisons on 1,2-anisyl shifts for the three systems shown in Table 6. In the systems being compared, the migrating group is the same in each case, but there are differences in the non-migrating aryl substituents at the migration origin or migration terminus.

It may be noted that transition states 34 and 32 differ only in the non-migrating An or Ph substituent initially present at the migration origin. The greater extent of rearrangement observed

TABLE 6

Comparison on 1,2-Anisyl Shifts in Reactions with HOAc -AgOAc

Substrate	Vinyl Cation	Transition State	% Rearrangement	k_r/k_{SOH} [a]	Ref.
12	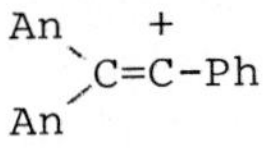 An C=C-Ph An 33	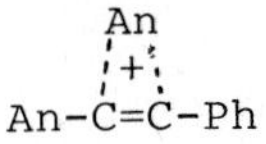 An + An-C=C-Ph 34	100	large	71
11a-2-Ph-d^5	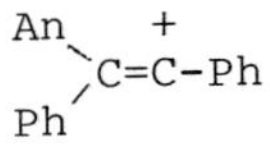 An C=C-Ph Ph 31	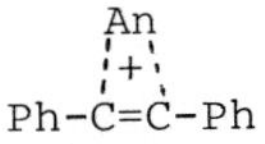 An + Ph-C=C-Ph 32	90	6.3	72
19a-2-^{13}C	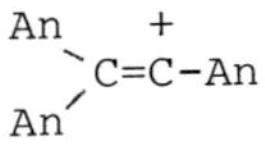 An C=C-An An 35	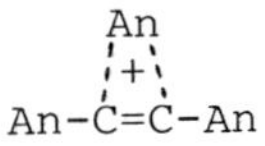 An + An-C=C-An 36	40	0.33	51

[a] Calculated using eq. 6.

for ion 33 than 31 may reflect the higher charge stabilization by the non-migrating An substituent in the bridged transition state 34. For transition states 34 and 36, the different An and Ph substituents are present initially at the migration terminus. The greater extent of rearrangement and larger k_r/k_{SOH} for 33 over 35 may be attributable to a faster capture reaction of the less stable α-phenylvinyl cation 33 than the α-anisylvinyl cation 35, as well as to the higher electrophilicity of the α-carbon in 33 than 35. In the latter case, the positive charge may be more extensively delocalized over the α-anisyl substituent.[65,66] Thus it was concluded that the higher charge stabilizing ability of an anisyl group when compared with a phenyl group is manifested whether these groups are at the migration origin, at the migration terminus, or are the migrating groups themselves.

The above conclusion may be viewed as in support of the finding that solvolyses with 2-anisyl-1,2-diphenylvinyl systems, such as cis- and trans-11a, proceed to the dissociated ion stage, while solvolyses with triphenylvinyl systems give rise to ion-pairs (Section II.B.2). Although both of these systems produce α-phenylvinyl cations, the 2-anisyl substituent can exert a stabilizing effect, resulting in the formation of the dissociated 2-anisyl-1,2-diphenylvinyl cation **31**.

VI. THE 1,2-DIANISYL-2-PHENYLVINYL SYSTEM

cis- and trans-1,2-Dianisyl-2-phenylvinyl bromides and chlorides (**37a** and **37b**) were prepared and their stereochemistry determined by Rappoport and Apeloig in 1969.[75] Solvolytic studies in various solvents, with or without added nucleophiles, were found to give a 1:1 mixture of the cis- and trans-products. Further work with cis- and trans-**37a**, cis-**37b** as well as the mesylates cis- and trans-**37c**, including kinetic studies and cis-trans isomerization in the recovered

$$\begin{array}{ccc} Ph & \diagdown & X \\ & C{=}C & \\ An \diagup & & \diagdown An \\ \end{array} \qquad X \;=\; Br \quad Cl \quad CH_3SO_3 \quad OAc$$

$$\underline{37} \hspace{4.5cm} \underline{a} \quad \underline{b} \qquad \underline{c} \qquad \underline{d}$$

reactant, were reported in 1975.[58] The results were interpreted by a mechanism involving the formation of an ion-pair which gives internal return with isomerization, or ionizes further to dissociated ions which may either give external ion return or solvolysis product, and this mechanism was verified by a computer simulation method.

A. Studies with labeled cis- and trans-1,2-dianisyl-2-phenylvinyl

bromides

In 1976, Lee and Oka[76] reported that the reaction of either cis- or trans-1,2-dianisyl-2-phenyl[2-^{13}C]vinyl bromide (cis- or trans-**37a**-2-^{13}C) with HOAc-AgOAc gave a 1:1 mixture of cis- and trans-1,2-dianisyl-2-phenyl[2-^{13}C]vinyl acetates (cis- and trans-**37d**-2-^{13}C) with no isotopic scrambling, indicating no degenerate rearrangement in the 1,2-dianisyl-2-phenylvinyl cation. In conjunction with their studies with 1,2-dianisyl-2-p-($^2H_3$)methoxyphenylvinyl bromide (**19a**-2-methoxy-d_3) (Section III.C.1), Rappoport et al.[56] also investigated the possibility of degenerate rearrangements during solvolyses with a 5:1 mixture of cis- and trans-1-anisyl-2-p-($^2H_3$) methoxyphenyl-2-phenylvinyl bromides (cis- and trans-**37a**-2-methoxy-d_3) as substrate. The results are summarized in Table 7.

From Table 7, it is seen that in the acetolysis, there was no detectable scrambling within the limit of measurement by ^{1}H n.m.r., and this finding is in agreement with the observation of Lee and Oka.[76] As stated earlier (Section III.C.1), the best solvent in promoting rearrangement was TFE. Rappoport et al.[56] also considered the relative ease of the 1,2-phenyl and 1,2-anisyl shifts in the various

TABLE 7

Degenerate Rearrangements During Solvolyses of 5:1 Mixtures of
cis- and trans-37a-2-methoxy-d$_3$

Solvent	Added salt	% Rearrangement	
		From n.m.r.	From m.s.
60% EtOH	2,6-lutidine	8 ± 2	$\underline{5\pm1}$
80% EtOH	2,6-lutidine	28 ± 6	$\underline{4\pm1.5}$
HOAc	NaOAc	$\underline{<3}^b$	7 ± 2
HOAc	NaOAc^c	$\underline{<3}^{b,d}$	
(CH$_3$)$_3$CCOOH	(CH$_3$)$_3$CCOONa	$\underline{0}^{d,e}$	
TFE	2,6-lutidine	53 ± 6	

^{a}Complete 100% rearrangement corresponds to the formation of 50% of
the scrambled 2-anisyl-1-p-(^{2}H$_3$)methoxyphenyl-2-phenylvinyl product;
the more reliable values are underlined.

^{b}Within the limits of detection by the n.m.r. method.

^{c}With the presence of added Bu$_4$NBr.

^{d}In the recovered reactant

^{e}No solvolytic substitution product; the recovered reactant was a
1:1 mixture of the cis- and trans-bromides.

possible triarylvinyl cations derived from different combinations of
phenyl and anisyl groups as the aryl substituents. With the same
combination of Ph and/or An substituents at the migration origin and
migration terminus, the An migration is, of course, favored over Ph
migration. When the aryl substituents at the migration origin and/or
migration terminus are different, migration to an α-phenyl substituted
cationic center is favored over the corresponding migration to an
α-anisylvinyl cation. Thus the non-degenerate migration of An from an
anisyl substituted migration origin to a phenyl substituted migration
terminus is the most favored process (eq. 16).

$$\begin{matrix} An \\ \diagdown \\ C=C-Ph \\ \diagup \\ An \end{matrix} \xrightarrow{} \begin{matrix} +An \\ An-C=C \\ \diagdown \\ Ph \end{matrix} \tag{16}$$

On the other hand, rearrangement of an α-anisylvinyl cation to an
α-phenylvinyl cation (eqs. 17 and 18, the latter being the reverse
of eq. 16) has never been observed.

$$\begin{matrix} Ph \\ \diagdown \\ {}^{+} \\ Ph \diagup \end{matrix} C=C-An \quad \xrightarrow{\ //\ } \quad Ph-C=C \begin{matrix} {}^{+}\ \diagup Ph \\ \diagdown \\ An \end{matrix} \tag{17}$$

$$\begin{matrix} An \\ \diagdown \\ {}^{+} \\ Ph \diagup \end{matrix} C=C-An \quad \xrightarrow{\ //\ } \quad Ph-C=C \begin{matrix} {}^{+}\ \diagup An \\ \diagdown \\ An \end{matrix} \tag{18}$$

Among the degenerate rearrangements, the process being discussed in this section, involving Ph migration between an An substituted migration origin, and an An substituted migration terminus, is the least favorable case.

Possible comparisons between the ease of 1,2-aryl shifts in triarylvinyl and triarylethyl systems were also attempted.[56] It was considered that only one pair of data would appear to be sufficiently closely related for a valid comparison. The acetolysis of 1,2,2-triphenyl[1-^{14}C]ethyl tosylate gave a 39% Ph migration from C-1 to C-2,[77] while a similar acetolysis of triphenyl[2-^{14}C]vinyl triflate[20] gave 6.7% scrambling (Section II.A.). Applying eq. 6, $k_{SOH}/k_{r(Ph)}$ can be calculated as 0.56 and 13, respectively, for the triphenylethyl tosylate and triphenylvinyl triflate. If k_{SOH} were assumed to be about the same (although this may not necessarily be the case), then the ease of 1,2-phenyl shift in the saturated triphenylethyl cation would be about 13/0.56 or 23 times as fast as the corresponding 1,2-phenyl shift in the triphenylvinyl system.

VII. THE 2-PHENYL-1,2-DI-p-TOLYLVINYL SYSTEM

A. Studies with cis- and trans-2-phenyl-1,2-di-p-tolyl[2-^{13}C]vinyl bromides (cis- and trans-38a-2-^{13}C)

A recent study in this laboratory dealt with the degenerate rearrangement from 1,2-phenyl shifts in the 2-phenyl-1,2-di-p-tolylvinyl system.[78] The reaction of cis- or trans-38a with HOAc containing 1.1 equiv. of AgOAc gave a 1:1 mixture of cis- and trans-2-phenyl-1,2-di-p-tolylvinyl acetates (cis- and trans-38b). The rate of solvolysis of either cis- or trans-38a in HOAc containing 2.0 equiv. of NaOAc was essentially the same and there was no significant solvent isotope effect for reactions in HOAc or DOAc. On the other hand, the solvolysis of cis-38a or a 1:1 mixture of cis- and trans-38a in unbuffered CF$_3$COOH or CF$_3$COOD gave a solvent isotope effect, $k_{CF_3COOH}/k_{CF_3COOD}$, of 3.4-3.9. The results of the degenerate rearrangement studies with cis- and trans-38a-2-^{13}C

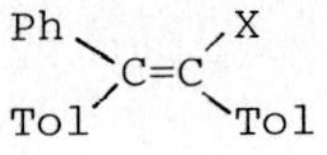

Ph   X   X = Br OAc CF_3COO
  C=C
Tol   Tol   a   b   c
  38

are summarized in Table 8.

TABLE 8

Scrambling Data from Solvolyses of cis- and trans-2-Phenyl-1,2-di-p-tolyl[2-^{13}C]vinyl Bromides (cis- and trans-38a-2-^{13}C)

Substrate[a]	Solvent and added salt	Reaction time, min.	Temp.	% Scrambling from C-2 to C-1
cis	HOAc–AgOAc	180	Reflux	2.0
trans	HOAc–AgOAc	180	Reflux	1.5
cis–trans	HOAc–AgOAc	1200	Reflux	1.5
cis	CF_3COOH–CF_3COOAg	20	Room	35
trans	CF_3COOH–CF_3COOAg	20	Room	34
cis–trans	CF_3COOH–CF_3COOAg	180	Room	35
cis–trans	CF_3COOH	105[b]	100°	45
cis–trans	CF_3COOH	260[c]	100°	48; 49

[a] The substrates were pure cis- or trans-38a-2-^{13}C or a 3:2 mixture of the cis and trans isomers.

[b] About 2.5 half-lives.

[c] About 6 half-lives; longer heating gave only decomposition.

From Table 8 it is seen that nearly identical extents of scrambling were observed in the reaction of cis- or trans-38a-2-^{13}C with HOAc–AgOAc or with CF_3COOH–CF_3COOAg. These results, together with the similarity in their rates of acetolysis and the formation of equal amounts of cis- and trans-product from either cis- or trans-reactant, all point to the formation of the dissociated, linear 2-phenyl-1,2-di-p-tolylvinyl cation in these reactions.

In the unbuffered trifluoroacetolysis of 38a or 38a-2-^{13}C, the behaviors observed are very similar to those found for tri-p-tolylvinyl bromide(26a or 26a-2-^{13}C). As discussed in Section IV.A, the significant solvent isotope effect (3.4–3.9) indicates an important role for electrophilic addition-elimination. The increase in the extent of scrambling with an increase in reaction time (Table 8), although small, also suggests that the initial product, 38c, might ionize and scramble further upon being heated in the reaction medium. On the other hand, the reaction of $\underline{38a}$-2-^{13}C with CF_3COOH-CF_3COOAg showed greatly enhanced rate and the product was stable at the room temperature condition that was used in the experiment. With Ag^+ catalysis, therefore, the trifluoroacetolysis probably proceeded via the S_N1 mechanism, without any significant complication by the slower electrophilic addition-elimination.

The reaction of $\underline{cis}$- or $\underline{trans}$-$\underline{38a}$-2-^{13}C with HOAc-AgOAc gave 1.5-2.0% scrambling. This value may be compared with 6-7% scrambling and no scrambling, respectively, for similar reactions of triphenyl[2-^{13}C]vinyl bromide ($\underline{10g}$-2-^{13}C) (Section II.B.1) and 1,2-dianisyl-2-phenyl[2-^{13}C]vinyl bromide ($\underline{37a}$-2-^{13}C) (Section VI.A). This comparison shows that, for the degenerate rearrangements, in going from the initial open triarylvinyl cations to the corresponding bridged transition states, the activation energies follow the order $\underline{41} \rightarrow \underline{44} > \underline{40} \rightarrow \underline{43} > \underline{39} \rightarrow \underline{40}$. The orders of

$$\begin{array}{ccc}
\text{Ph} \quad + & \text{Ph} \quad + & \text{Ph} \quad + \\
\text{C=C-Ph} & \text{C=C-Tol} & \text{C=C-An} \\
\text{Ph} & \text{Tol} & \text{An} \\
\underline{39} & \underline{40} & \underline{41}
\end{array}$$

$$\begin{array}{ccc}
\text{Ph} & \text{Ph} & \text{Ph} \\
/+\backslash & /+\backslash & /+\backslash \\
\text{Ph-C=C-Ph} & \text{Tol-C=C-Tol} & \text{An-C=C-An} \\
\underline{42} & \underline{43} & \underline{44}
\end{array}$$

stability are expected to be $\underline{41} > \underline{40} > \underline{39}$ and $\underline{44} > \underline{43} > \underline{42}$. Apparently the stabilizing effect of the α-aryl substituent on the vinyl cation is greater than the stabilization of the bridged transition state by the non-migrating aryl groups. Hence the energy is lowered to a greater extent in the initial vinyl cation than in the transition state, and this effect is greatest with an α-anisyl substituent, thus giving rise to the observed order of activation energies, namely, $\underline{41} \rightarrow \underline{44} > \underline{40} \rightarrow \underline{43} > \underline{39} \rightarrow \underline{42}$.

Of the various degenerate rearrangements in the different triarylvinyl cationic systems reviewed in this chapter, except for the triphenylvinyl case, all reactions were shown to proceed via the dissociated triarylvinyl cation. Under solvolytic conditions, 1,2-phenyl shifts in the triphenylvinyl system apparently took place at the ion-pair stage. It is of interest to note that from a comprehensive solvolytic study with a series of $\underline{cis}$- and $\underline{trans}$-2-

aryl-1,2-dimethylvinyl triflates (45), definitive evidence in
support of aryl participation with anchimeric assistance has been
obtained by Stang and Dueber.[79,80] For example, in the solvolysis
of trans-1,2-dimethyl-2-phenylvinyl triflate, the phenyl bridged
ion 46 was shown to be the reaction intermediate. Incidentally,

Ar   OTf
  \   /
   C=C
  /   \
CH₃   CH₃   CH₃-C=C-CH₃

45   46

these workers[79,80] referred to 46 as a vinylidene phenonium ion.
Since the divalent vinylidene radical is $CH_2=C=$, and $-CH=CH-$ is the
vinylene radical, perhaps 46 may be more properly termed a vinylene
phenonium ion, or if the nomenclature proposed by Olah[81] is adopted,
46 specifically should be named as 2,3-butenylenebenzenium ion.

   Since an α-methyl substituted vinyl cation may be less stable than
an aryl bridged ion such as 46, solvolysis with 2-aryl-1,2-dimethyl-
vinyl substrates would involve aryl participation with anchimeric
assistance, as was observed by Stang and Dueber. In the triarylvinyl
systems, however, the stability of an α-tolyl or α-anisyl substituted
vinyl cation apparently is sufficiently high so that no anchimeric
assistance from β-aryl participation is needed and ionization proceeds
to the dissociated ion. The triphenylvinyl cation is the least stable
of the triarylvinyl cations that have been studied, and its ionization
may not proceed as far as the other more stable triarylvinyl cations.
Thus, 1,2-shifts and capture by solvent can occur at the ion-pair
stage for the triphenylvinyl system, and this behavior could be
regarded as falling into the borderline region between the α-methyl
and α-tolyl or α-anisyl substituted vinyl cations.

   It was pointed out, in Section IV.A, that in evaluating migratory
aptitudes in triarylvinyl cationic systems, comparisons should be
made with data from systems having different migrating groups but
with the same non-migrating substituents at the migration origin
and migration terminus. In calculating the migratory aptitudes
using the rearrangement data from the 2-anisyl-1,2-diphenylvinyl
system, as described in Section V.A, the corrected values of 40-63
obtained for the An/Ph migration ratio were based on competing
migrations in the 2-anisyl-1,2-diphenylvinyl cation (31). The
results, therefore, reflect the differences in the energies of
transition states 32 and 47, in which the two non-migrating aryl

$$\begin{array}{ccc}
\underset{Ph}{\overset{An}{>}}C=C-Ph & \overset{An}{\underset{}{Ph-C=C-Ph}} & \overset{Ph}{\underset{}{An-C=C-Ph}} \\
\underline{31} & \underline{32} & \underline{47}
\end{array}$$

substituents are not the same.

Rappoport et al. have also compared the scrambling data from reactions of HOAc-AgOAc with trans-2-anisyl-2-(^{2}H$_5$)phenyl-1-phenylvinyl bromide (trans-lla-2-Ph-d$_5$)[72] and with triphenyl[2-^{13}C]vinyl bromide (10g-2-^{13}C).[35] With the observed degenerate rearrangement of 92.7 ± 1.8% for trans-lla-2-Ph-d$_5$ and the ^{13}C scrambling of 6-7% for 10g-2-^{13}C, using eq. 6, $k'_{SOH}/k_{r(An)}$ and $k_{SOH}/k_{r(Ph)}$ were calculated to be 0.16 and 13, respectively. Making the reasonable assumption that k'_{SOH} and k_{SOH} in HOAc-AgOAc for the two substrates are about equal, a value of 13/0.16 or 81 for the migration ratio of $k_{r(An)}/k_{r(Ph)}$ was obtained.[72] In the same way, the 1.5-2.0% scrambling obtained in the reaction of cis- or trans-2-phenyl-1,2-di-p-tolyl-[2-^{13}C]vinyl bromide (cis- or trans-38a-2-^{13}C) with HOAc-AgOAc[78] may be compared with the 13-14% scrambling observed in a similar reaction with tri-p-tolyl[2-^{13}C]vinyl bromide (26a-2-^{13}C).[60] From eq. 6 and using mean values of 1.7 and 13.5% scramblings, respectively, for the reactions of 38a-2-^{13}C and 26a-2-^{13}C, the $k_{SOH}/k_{r(Ph)}$ and $k'_{SOH}/k_{r(Tol)}$ values would be 57 and 5.4. Again assuming k_{SOH} and k'_{SOH} to be the same, the migration ratio for $k_{r(Tol)}/k_{r(Ph)}$ is about 11.

The above treatments have the advantage that the data used for comparison were derived from systems that differed only in the migrating group, with identical aryl substituents at the migration origin and migration terminus, i.e. An migration in 31 compared with Ph migration in 39, and Ph migration in 40 compared with Tol migration in 48. As has been pointed out in the literature,[56,72]

$$\begin{array}{cccc}
\underset{Ph}{\overset{An}{>}}C=C-Ph & \underset{Ph}{\overset{Ph}{>}}C=C-Ph & \underset{Tol}{\overset{Ph}{>}}C=C-Tol & \underset{Tol}{\overset{Tol}{>}}C=C-Tol \\
\underline{31} & \underline{39} & \underline{40} & \underline{48}
\end{array}$$

in calculations using eq. 6, when the extent of scrambling is either very large or very small, such as in the reaction with cis-lla-2-Ph-d$_5$ or with 38a-2-^{13}C, slight differences in the extent of scrambling used in the calculation will result in large variations in k_{SOH}/k_r. Thus the migration ratios obtained are subjected to large errors.

Nevertheless, using the mean scrambling values for the calculation, the data indicated that the relative migratory aptitudes for Ph:Tol: An should be about 1:11:81. These values, however, have to be modified as discussed below.

In using eq. 6 to evaluate migratory aptitudes, an omission has been made both by Rappoport[72] and by us.[78] Two variations of Scheme 1 (page 9) are shown in Schemes 6 and 7. It is seen that

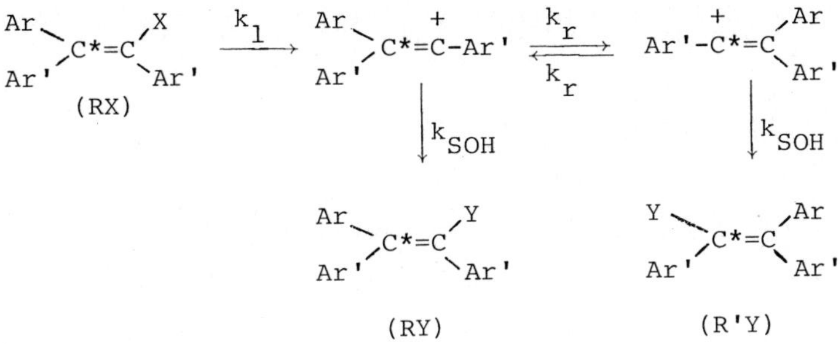

$$\begin{array}{l}\underset{Ar'}{\overset{Ar}{>}}C*=C\underset{Ar'}{\overset{X}{<}} \quad \xrightarrow{k_1} \quad \underset{Ar'}{\overset{Ar}{>}}C*=C-Ar' \quad \underset{k_r}{\overset{k_r}{\rightleftharpoons}} \quad Ar'-C*=C\underset{Ar'}{\overset{Ar}{<}} \end{array}$$

(RX)

$$\downarrow k_{SOH} \qquad\qquad\qquad \downarrow k_{SOH}$$

$$\underset{Ar'}{\overset{Ar}{>}}C*=C\underset{Ar'}{\overset{Y}{<}} \qquad\qquad \underset{Ar'}{\overset{Y}{>}}C*=C\underset{Ar'}{\overset{Ar}{<}}$$

(RY)   (R'Y)

SCHEME 6

$$\begin{array}{l}\underset{Ar}{\overset{Ar}{>}}C*=C\underset{Ar}{\overset{X}{<}} \quad \xrightarrow{k_1} \quad \underset{Ar}{\overset{Ar}{>}}C*=C-Ar \quad \underset{2k_r}{\overset{2k_r}{\rightleftharpoons}} \quad Ar-C*=C\underset{Ar}{\overset{Ar}{<}} \end{array}$$

$$\downarrow k_{SOH} \qquad\qquad\qquad \downarrow k_{SOH}$$

$$\underset{Ar}{\overset{Ar}{>}}C*=C\underset{Ar}{\overset{Y}{<}} \qquad\qquad \underset{Ar}{\overset{Y}{>}}C*=C\underset{Ar}{\overset{Ar}{<}}$$

(RY)   (R'Y)

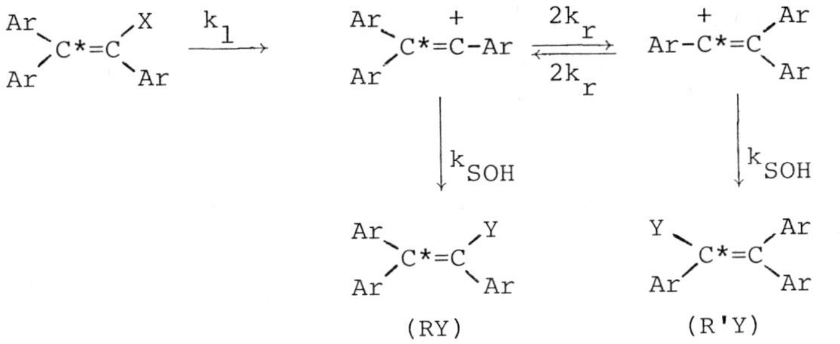

SCHEME 7

in Scheme 7, a statistical factor of 2 has to be employed. In Scheme 6, there is no difference from Scheme 1, and eq. 6, i.e. $[RY]/[R'Y] = 1 + (k_{SOH}/k_r)$, still applies. For Scheme 7, however, integration gave $[RY]/[R'Y] = 1 + (k_{SOH}/2k_r)$. Hence in comparing the data from the 2-anisyl-1,2-diphenylvinyl system cis-11a-2-Ph-d_5 and from the triphenylvinyl system 10g-2-^{13}C, as given on page 38, $k_{SOH}/k_{r(An)}$ is still 0.16, but $k_{SOH}/2k_{r(Ph)}$ is 13, and $k_{r(An)}/k_{r(Ph)}$ should be 2 x 81 or 162. Similarly, in using the data from the 2-phenyl-1,2-di-p-tolylvinyl system 38a-2-^{13}C and from the tri-p-tolylvinyl system 26a-2-^{13}C, $k_{SOH}/k_{r(Ph)}$ is still 57, but $k_{SOH}/2k_{r(Tol)}$ is 5.4, giving a ratio for $k_{r(Tol)}/k_{r(Ph)}$ of about 5. Thus the corrected values for the migratory aptitudes of Ph:Tol:An derived from reactions of triarylvinyl bromides with HOAc-AgOAc should be about 1:5:182, and these values are certainly increasing in the order expected.

VIII. CONCLUDING REMARKS

As stated in the introductory Section I, in the detection of degenerate rearrangements in triarylvinyl cations, it is necessary to use isotopes as labels. In fact, the data reviewed in this chapter would not have been obtainable without employing such labels. The results obtained, however, have all been explained by utilizing theories and interpretations similar to those that have already been applied in explaining the behaviors of ordinary saturated carbocations. These findings, therefore, support the view that vinyl cations, once formed, would behave generally in the same way as other carbocations, although in specific cases, special behaviors may be observed, such as the involvement of ion-pairs in the triphenylvinyl system and the high selectivity of α-anisyl substituted triarylvinyl cations, possibly because of a steric effect.

With the phenyl and anisyl groups as the aryl substituents, some work has been done on all of the four possible degenerate rearrangements involving the triphenylvinyl, trianisylvinyl, 2-anisyl-1,2-diphenylvinyl and 1,2-dianisyl-2-phenylvinyl cationic systems. With the introduction of the p-tolyl group as another variable aryl substituent, besides the work already done on the tri-p-tolylvinyl and 2-phenyl-1,2-di-p-tolylvinyl systems, further investigations of degenerate rearrangements may be extended to include the 1,2-diphenyl-2-p-tolylvinyl, 2-anisyl-1,2-di-p-tolylvinyl and 1,2-dianisyl-2-p-tolylvinyl systems. It may be questionable whether more, essentially similar, experiments will be capable of providing any novel results. However, further work using one or more of the three additional systems just listed should at least provide further data to confirm the relative migratory aptitudes that have already been evaluated for the phenyl, p-tolyl and anisyl groups in triarylvinyl cations.

All aryl groups utilized so far in studies of triarylvinyl cations exert stabilizing effects on these cations. It could be of some significance if the work reviewed in this chapter were to be extended to include triarylvinyl reactants with destabilizing aryl substituents. For example, with a 2-aryl-1,2-di-m-methoxyphenylvinyl substrate, it would be of interest to ascertain, in such a vinyl cation of relatively low stability, whether β-aryl participation could occur with anchimeric assistance and with the formation of a bridged ion as reaction intermediate. One might also anticipate that in such a system, the rate of reaction may become too slow. However, a change to the triflate instead of the bromide as a leaving group may alleviate such problems. Finally, it may be indicated that collaborative studies between Professor Z. Rappoport and us, using doubly labelled triarylvinyl substrates, are currently underway. It is anticipated, for example, that with such double labelling, data may be obtained which might show just how extensive in the possibility of repeated, back and forth 1,2-aryl shifts in triarylvinyl cationic systems.

REFERENCES

1   C. A. Grob and G. Cseh, Helv. Chim. Acta, 47 (1964) 194.
2   R. T. Morrison and R. N. Boyd, Organic Chemistry, Allyn and Bacon, Inc., Boston, Mass., 2nd Ed., 1966, p. 828; 3rd Ed., 1973, p. 823.
3   J. D. Roberts and M. C. Caserio, Basic Principles of Organic Chemistry, W. A. Benjamin, Inc., New York, N.Y., 1964, p. 321. In the 2nd edition, 1977, p. 549, it is pointed out that vinyl cations may be generated with substrates containing a superior leaving group, such as the triflate.
4   M. Hanack, Accounts Chem. Res. 3 (1970) 209; 9 (1976) 364.
5   C. A. Grob, Chimia, 25 (1971) 87.
6   G. Modena and U. Tonellato, Adv. Phys. Org. Chem., 9 (1971) 185.
7   P. J. Stang, Prog. Phys. Org. Chem., 10 (1973) 205.
8   M. Hanack and L. R. Subramanian, J. Chem. Educ., 52 (1975) 80.
9   Z. Rappoport, Accounts Chem. Res., 9 (1976) 265.
10 G. Modena and U. Tonellato, Chem. Comm., (1968) 1676; J. Chem. Soc. B, (1971) 374.
11 G. Capozzi, G. Melloni, G. Modena and U. Tonellato, Chem. Comm., (1969) 1520.
12 W. M. Jones and F. W. Miller, J. Am. Chem. Soc., 89 (1967) 1960.
13 M. A. Imhoff, R. H. Summerville, P. v. R. Schleyer, A. G. Martinez, M. Hanack, T. E. Dueber, and P. J. Stang, J. Am. Chem. Soc., 92 (1970) 3802.
14 L. L. Miller and D. A. Kaufman, J. Am. Chem. Soc., 90 (1968) 7282.
15 W. M. Jones and D. D. Maness, J. Am. Chem. Soc., 91 (1969) 4341; 92 (1970) 5457.
16 V. J. Shiner and J. G. Jewett, J. Am. Chem. Soc., 86 (1964) 945.
17 D. S. Noyce and D. Schiavelli, J. Am. Chem. Soc., 90 (1968) 1023.

18 P. J. Stang and R. Summerville, J. Am. Chem. Soc., 91 (1969) 4600.

19 Z. Rappoport and Y. Houminer, J. Chem. Soc. Perkin II, (1973)
    1506.

20 C. C. Lee, A. J. Cessna, B. A. Davis and M. Oka, Can. J. Chem.,
    52 (1974) 2679.

21 W. G. Young, S. Winstein and H. L. Goering, J. Am. Chem. Soc.,
    73 (1951) 1958.

22 A. Streitwieser, Jr., Solvolytic Displacement Reactions, McGraw-
    Hill, New York, N.Y., 1962, pp. 86-88.

23 C. C. Lee and E. C. F. Ko, Can. J. Chem., 54 (1976) 3041.

24 C. C. Lee, G. P. Slater and J. W. T. Spinks, Can. J. Chem.,
    35 (1957) 1417.

25 J. E. Norlander and W. G. Deadman, J. Am. Chem. Soc., 90 (1968)
    1590.

26 J. L. Coke, F. E. McFarlane, M. C. Mourning and M. G. Jones,
    J. Am. Chem. Soc., 91 (1969) 1154.

27 C. C. Lee and D. Unger, Can. J. Chem., 52 (1974) 3955.

28 C. J. Lancelot, D. J. Cram and P. v. R. Schleyer, In G. A. Olah
    and P. v. R. Schleyer, Editors, Carbonium Ions, Volume III,
    Wiley-Interscience, New York, N.Y., 1972, Chap. 27.

29 D. R. Kelsey and R. G. Bergman, J. Am. Chem. Soc., 93 (1971)
    1953.

30 T. C. Clarke, D. R. Kelsey and R. G. Bergman, J. Am. Chem. Soc.,
    94 (1972) 3626; T. C. Clarke and R. G. Bergman, J. Am. Chem. Soc.,
    94 (1972) 3627; 96 (1974) 7934.

31 R. H. Summerville and P. v. R. Schleyer, J. Am. Chem. Soc., 94
    (1972) 3629; 96 (1974) 1110.

32 A. H. Fainberg and S. Winstein, J. Am. Chem. Soc., 78 (1956)
    2770.

33 P. E. Peterson and F. J. Waller, J. Am. Chem. Soc., 94 (1972) 991.

34 T. W. Bentley, F. L. Schadt and P. v. R. Schleyer, J. Am. Chem.
    Soc., 94 (1972) 992; F. L. Shadt, T. W. Bentley and P. v. R.
    Schleyer, J. Am. Chem. Soc., 98 (1976) 7667.

35 F. H. A. Rummens, R. D. Green, A. J. Cessna, M. Oka and C. C. Lee,
    Can. J. Chem., 53 (1975) 314.

36 Z. Rappoport, private communications.

37 G. F. P. Kernaghan and H. M. R. Hoffmann, J. Am. Chem. Soc., 92
    (1970) 6988.

38 C. C. Lee and E. C. F. Ko, Can. J. Chem., 56 (1978) 2459.

39 Unpublished results of E. C. F. Ko.

40 C. J. Collins and W. A. Bonner, J. Am. Chem. Soc., 77 (1955) 92;

W. A. Bonner and C. J. Collins, J. Am. Chem. Soc., 78 (1956) 5587.

41   C. C. Lee, E. C. F. Ko and Z. Rappoport, to be published.

42   S. Winstein, E. Clippinger, A. H. Fainberg, R. Heck and G. C. Robinson, J. Am. Chem. Soc., 78 (1956) 328.

43   M. Hanack, T. Bässler, W. Eymann, W. E. Heyd and R. Kopp, J. Am. Chem. Soc., 97 (1974) 6686.

44   C. C. Lee, R. Tkachuk and G. P. Slater, Tetrahedron, 7 (1959) 206.

45   J. W. Clayton and C. C. Lee, Can. J. Chem., 39 (1961) 1510.

46   Z. Rappoport and A. Gal, J. Am. Chem. Soc., 91 (1969) 5246.

47   Z. Rappoport and Y. Apeloig, Tetrahedron Lett., (1970) 1817.

48   Z. Rappoport and A. Gal, Tetrahedron Lett., (1970) 3233.

49   Z. Rappoport and J. Kaspi, J. Am. Chem. Soc., 92 (1970) 3220; J. Chem. Soc. Perkin II, (1972) 1102.

50   T. Sonoda, M. Kawakami, T. Ikeda, S. Kobayashi and H. Taniguchi, J. C. S. Chem. Comm., (1976) 612.

51   M. Oka and C. C. Lee, Can. J. Chem., 53 (1975) 320.

52   J. Hinton, M. Oka and A. Fry, In E. Buncel and C. C. Lee, Editors, Isotopes in Organic Chemistry, Volume 3, Carbon-13 in Organic Chemistry, Elsevier Scientific Publishers, Amsterdam, 1977, Chap. 2.

53   M. Oka, Ph.D. Thesis, University of Arkansas, 1973.

54   U. Weber, Ph.D. Thesis, University of Saskatchewan, 1978.

55   Z. Rappoport, P. Shulman and M. Thuval (Shoolman), J. Am. Chem. Soc., 100 (1978) 7041.

56   Y. Houminer, E. Noy and Z. Rappoport, J. Am. Chem. Soc., 98 (1976) 5632.

57   J. Murto and C. L. Heino, Suomen. Kemistil. B, 39 (1966) 263.

58   Z. Rappoport and Y. Apeloig, J. Am. Chem. Soc., 97 (1975) 821; 97 (1975) 836.

59   C. C. Lee, U. Weber and C. A. Obafemi, Can. J. Chem., 57 (1979) 1384.

60   C. C. Lee, A. J. Paine and E. C. F. Ko, Can. J. Chem., 55 (1977) 2010.

61   W. E. Bachmann and J. W. Ferguson, J. Am. Chem. Soc., 56 (1934) 2081.

62   L. S. Ciereszko and J. G. Burr, Jr., J. Am. Chem. Soc., 74 (1952) 5431.

63   J. D. Roberts and C. M. Regan, J. Am. Chem. Soc., 75 (1953) 2069.

64   D. Y. Curtin and M. C. Crew, J. Am. Chem. Soc., 76 (1954) 3719.

65   W. A. Bonner and T. A. Putkey, J. Org. Chem., 27 (1962) 2348.

66   M. J. McCall, J. M. Townsend and W. A. Bonner, J. Am. Chem. Soc.,

97 (1975) 2743.

67 P. E. Peterson and J. M. Indelicato, J. Am. Chem. Soc., 90 (1968)
    6515.

68 W. M. Schubert and G. W. Barfknecht, J. Am. Chem. Soc., 92 (1970)
    207.

69 Z. Rappoport, T. Bässler and M. Hanack, J. Am. Chem. Soc., 92
    (1970) 4985.

70 Z. Rappoport and A. Gal, J. Chem. Soc. Perkin II, (1973) 301.

71 Z. Rappoport, A. Gal and Y. Houminer, Tetrahedron Lett., (1973)
    641.

72 Z. Rappoport, E. Noy and Y. Houminer, J. Am. Chem. Soc., 98
    (1976) 2238.

73 W. E. Bachmann and F. H. Moser, J. Am. Chem. Soc., 54 (1932)
    1124.

74 J. G. Burr, Jr. and L. S. Ciereszko, J. Am. Chem. Soc., 74 (1952)
    5426.

75 Z. Rappoport and Y. Apeloig, J. Am. Chem. Soc., 91 (1969) 6734.

76 C. C. Lee and M. Oka, Can. J. Chem., 54 (1976) 604.

77 W. A. Bonner and C. J. Collins, J. Am. Chem. Soc., 75 (1953) 5372.

78 C. C. Lee, A. J. Paine and E. C. F. Ko, J. Am. Chem. Soc., 99
    (1977) 7267.

79 P. J. Stang and T. E. Dueber, J. Am. Chem. Soc., 95 (1973) 2683;
    2686.

80 P. J. Stang and T. E. Dueber, J. Am. Chem. Soc., 99 (1977) 2602.

81 G. Olah, J. Am. Chem. Soc. 94 (1972) 808.

CHAPTER 2

THE PUMMERER AND PUMMERER TYPE OF REACTIONS

Shigeru OAE and Tatsuo NUMATA
Department of Chemistry, University of Tsukuba, Sakura-mura,
Ibaraki, 300-31 Japan

I. INTRODUCTION

Sulfoxides bearing a methyl or methylene group directly attached to the sulfur atom can readily be transformed to α-substituted sulfides upon treatment with electrophilic reagents, the net result being the reduction of the sulfoxide group to the sulfide group and the oxidation of the α-carbon atom. When acetic anhydride is used, the reaction is illustrated in eq. 1.

$$R-\underset{\underset{O}{\downarrow}}{S}-CH_3 \ + \ (CH_3CO)_2O \ \longrightarrow \ R-S-CH_2-O-\underset{\underset{O}{\|}}{C}-CH_3 \ + \ CH_3COOH \quad (1)$$

$$\underline{1}$$

α-Acetoxymethyl sulfide ($\underline{1}$) is readily hydrolyzed upon dissolving in aqueous media to form the corresponding thiol, formaldehyde and acetic acid (eq. 2). The whole reaction proceeds so smoothly that

$$R-S-CH_2-O-\underset{\underset{O}{\|}}{C}-CH_3 \ \xrightarrow{\ H_2O\ } \ R-SH \ + \ CH_2O \ + \ CH_3COOH \quad (2)$$

$$\underline{1}$$

it has been suggested as a possible pathway for enzymatic demethylation of methionine.[1]

The first example of the Pummerer reaction was the acid-catalyzed decomposition of α-phenylsulfinylacetic acid ($\underline{2}$), found by Pummerer himself in 1909 to form either glyoxal and thiophenol (eq. 3), or α-chloro-α-phenylthioacetic acid ($\underline{3}$) when dry hydrogen chloride was used (eq. 4).[2] In the following year, 1910, Pummerer treated ethyl phenylsulfinylacetate ($\underline{4}$) with acetic anhydride and obtained ethyl α-phenylthio-α-acetoxyacetate ($\underline{5}$) (eq. 5). He also reported that treatment of α-phenylsulfinylpropionic acid ($\underline{6}$) with acid gave pyruvic

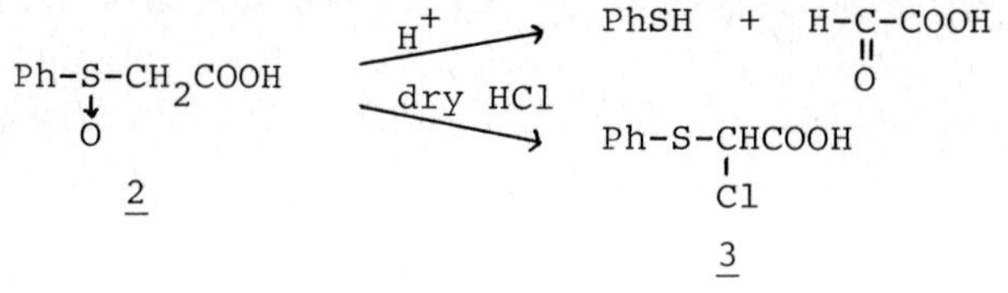

$$Ph\text{-}S\text{-}CH_2COOH \xrightarrow[\text{dry HCl}]{H^+}$$
$$\downarrow$$
$$O$$
$$\underline{2}$$

$$PhSH + H\text{-}C\text{-}COOH \qquad (3)$$
$$\qquad\qquad\quad \|$$
$$\qquad\qquad\quad O$$

$$Ph\text{-}S\text{-}CHCOOH \qquad\qquad (4)$$
$$\qquad\quad |$$
$$\qquad\quad Cl$$
$$\qquad\quad \underline{3}$$

acid and thiophenol (eq. 6).[3]

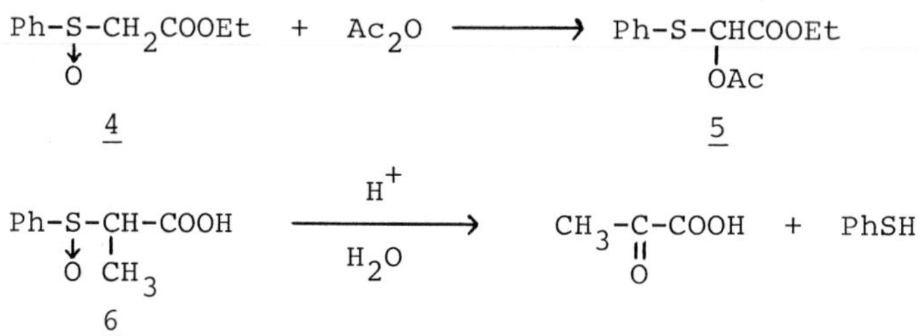

$$Ph\text{-}S\text{-}CH_2COOEt + Ac_2O \longrightarrow Ph\text{-}S\text{-}CHCOOEt \qquad (5)$$
$$\downarrow \qquad\qquad\qquad\qquad\qquad\qquad\qquad |$$
$$O \qquad\qquad\qquad\qquad\qquad\qquad\quad OAc$$
$$\underline{4} \qquad\qquad\qquad\qquad\qquad\qquad\qquad \underline{5}$$

$$Ph\text{-}S\text{-}CH\text{-}COOH \xrightarrow[H_2O]{H^+} CH_3\text{-}C\text{-}COOH + PhSH \qquad (6)$$
$$\downarrow \; | \qquad\qquad\qquad\qquad\quad \|$$
$$O \; CH_3 \qquad\qquad\qquad\qquad O$$
$$\underline{6}$$

The reaction was not reexamined until half a century later by Horner at the University of Mainz who reconfirmed that the acetic anhydride treatment of sulfoxides having α-methylene protons gives the corresponding α-acetoxysulfides in good yields.[4] He then proposed that the reaction in which a sulfoxide group is reduced to sulfide while an α-methylene group is oxidized simultaneously, such as that between the sulfoxide and an electrophile, like acetic anhydride, be called the Pummerer reaction.[5] Meanwhile, similar reactions have been found to take place with sulfimides and sulfonium ylides which are isoelectronic to sulfoxides. Halogenation at the α-methylene position of sulfides involving intermediary S-halosulfonium salts was also suggested by us[6] to follow similar mechanistic routes. Thus, the Pummerer reaction is no longer limited to that between sulfoxides and acid anhydrides.

We first investigated the mechanism of the Pummerer reaction of dimethyl sulfoxide and acetic anhydride around 1960 with the use of ^{18}O tracer and pointed out the intermolecular nature of the acetoxy migration via an ylide intermediate;[7] however we did not specify the involvement of α-sulfenyl-stabilized carbonium ion as an intermediate. The possible involvement of the carbonium ion intermediate was first suggested by Johnson in the following reaction (eq. 7).[8]

When an alkoxysulfonium salt was treated with base, two reactions were found to take place concurrently; the one is the Pummerer reaction and the other is the oxidative cleavage to give the corresponding carbonyl compound and the sulfide. Meanwhile a rapid alkoxy

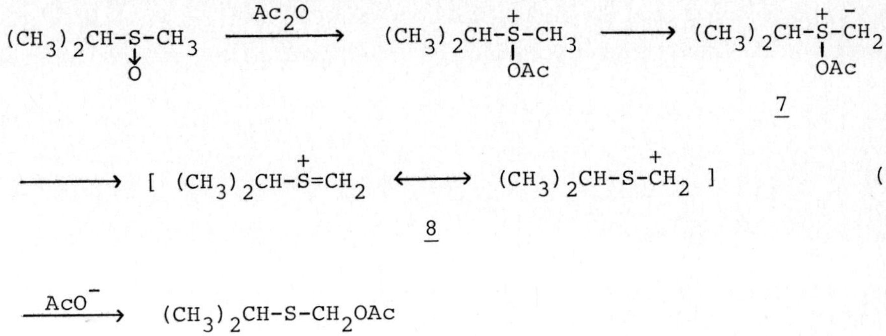

$$(CH_3)_2CH-\overset{\downarrow}{\underset{O}{S}}-CH_3 \xrightarrow{Ac_2O} (CH_3)_2CH-\overset{+}{\underset{OAc}{S}}-CH_3 \longrightarrow (CH_3)_2CH-\overset{+}{\underset{OAc}{S}}-\overset{-}{CH_2}$$

$$\underline{7}$$

$$\longrightarrow [\ (CH_3)_2CH-\overset{+}{S}=CH_2 \longleftrightarrow (CH_3)_2CH-S-\overset{+}{CH_2}\ ] \tag{7}$$

$$\underline{8}$$

$$\xrightarrow{AcO^-} (CH_3)_2CH-S-CH_2OAc$$

exchange usually takes place during the reaction. Benzylic methoxy-sulfonium salt undergoes preferentially the former reaction, whereas methyl phenyl methoxysulfonium salt follows the latter reaction predominantly (eq. 8).[9]

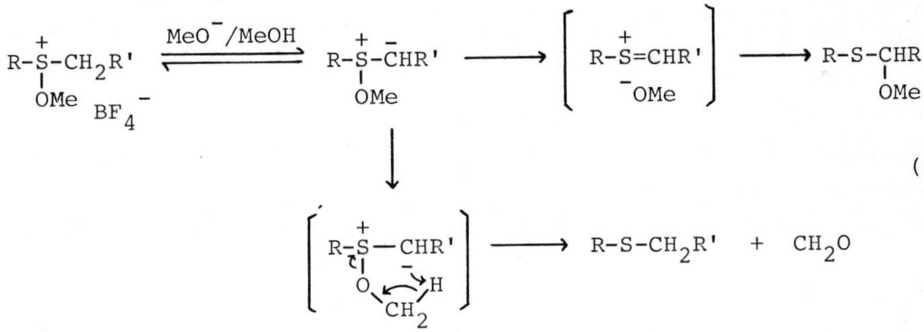

$$R-\overset{+}{\underset{\underset{BF_4^-}{OMe}}{S}}-CH_2R' \underset{}{\overset{MeO^-/MeOH}{\rightleftharpoons}} R-\overset{+}{\underset{OMe}{S}}-\overset{-}{C}HR' \longrightarrow \left[R-\overset{+}{\underset{OMe}{S}}=CHR' \right] \longrightarrow R-\underset{OMe}{S}-CHR'$$

$$\tag{8}$$

$$\left[R-\overset{+}{\underset{\overset{}{\underset{CH_2}{O\ \curvearrowright H}}}{S}}-CHR' \right] \longrightarrow R-S-CH_2R'\ +\ CH_2O$$

When the alkoxysulfonium salt was treated with sodium acetate in dimethyl sulfoxide, both the Pummerer reaction and the oxidation took place concurrently (eq. 9). Treatment with sodium propionate or p-nitrobenzoate leads to a similar reaction.[8,10]

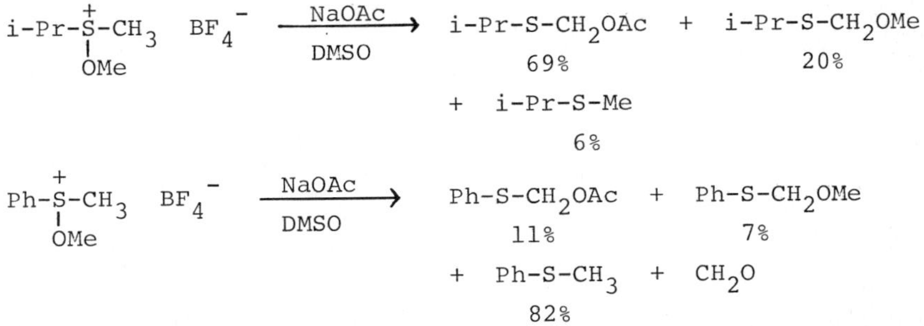

$$i-Pr-\overset{+}{\underset{OMe}{S}}-CH_3\ BF_4^- \xrightarrow[DMSO]{NaOAc} i-Pr-S-CH_2OAc\ +\ i-Pr-S-CH_2OMe$$
$$69\% \qquad\qquad 20\%$$
$$+\ i-Pr-S-Me$$
$$6\%$$

$$\tag{9}$$

$$Ph-\overset{+}{\underset{OMe}{S}}-CH_3\ BF_4^- \xrightarrow[DMSO]{NaOAc} Ph-S-CH_2OAc\ +\ Ph-S-CH_2OMe$$
$$11\% \qquad\qquad 7\%$$
$$+\ Ph-S-CH_3\ +\ CH_2O$$
$$82\%$$

The mechanism for these reactions does not deny the involvement of the sulfonium ylide intermediate (<u>7</u>), but it assumes the intervention

of an additional sulfur-stabilized carbonium ion intermediate (8)
prior to the rearrangement which is presumed to be intermolecular.
The intermediate carbonium ion (8) is no other than that formed in
the facile solvolysis of an α-chloroalkyl sulfide (9) (eq. 10).[6,11,12]

$$R-S-CH_2Cl \xrightarrow[\text{in protic solvent}]{\text{slow}} [\; R-\overset{+}{S}=CH_2 \longleftrightarrow R-S-\overset{+}{C}H_2 \;] \; + \; Cl^-$$

$$\underset{9}{} \qquad\qquad\qquad\qquad \underset{8}{}$$

$$\tag{10}$$

$$\xrightarrow[\text{H}_2\text{O}]{\text{fast}} \; R-S-CH_2OH \; + \; H^+ \xrightarrow[\text{H}_2\text{O}]{\text{fast}} \; R-SH \; + \; CH_2O$$

A sulfur-stabilized carbonium ion was isolated as the antimony
hexachloride salt (10) (which is hygroscopic and extremely reactive)
in the following treatment[13] (eq. 11) and shown to react with various

$$CH_3-S-CH_2Cl \; + \; SbCl_5 \xrightarrow{CH_2Cl_2} [\; CH_3-\overset{+}{S}=CH_2 \longleftrightarrow CH_3-S-\overset{+}{C}H_2 \;] \; SbCl_6^-$$

$$\underset{10}{} \qquad\qquad\qquad \tag{11}$$

nucleophiles at the α-carbon atom, the site of the carbonium ion
center.[21] Thus for nearly a decade, the Pummerer reaction was considered to involve the sulfur-stabilized carbonium ion intermediate.
A controversy arose when the first example of asymmetric induction
was found by Allenmark[14] and more convincing evidence for the intramolecular nature of the rearrangement was found with our [18]O tracer
experiment with α-cyanomethyl sulfoxide.[15] Thus, we felt it timely
to review the mechanism of the Pummerer reaction with the latest
information at this time, despite the numerous reviews which have
been published in recent years.[16-21]

II. GENERAL MECHANISM

Although there are numerous examples of the Pummerer reaction
between various sulfoxides and electrophilic reagents,[16,17] most
of the studies on the mechanism have been carried out with acetic
anhydride. Though the possible involvement of a sulfurane intermediate
is still being argued and while only a few of the suggested intermediates have actually been isolated, the general scheme of the
reaction is believed to be shown in the following four sequential
elemental reactions (eq. 12), with the rate-determining step varying

with changes of both the acylating agent and the sulfoxide.

$$R\text{-}S\text{-}CH_3 \ + \ (CH_3CO)_2O \ \xrightarrow{\text{step 1}} \ \overset{+}{R}\text{-}S\text{-}CH_3 \ + \ AcO^-$$
with O below the first S (downward arrow) and OAc below the product S.

$$\xrightarrow[-H^+]{\text{step 2}} \ \left[\overset{+}{R}\text{-}\overset{-}{S}\text{-}CH_2 \ \longleftrightarrow \ R\text{-}S\text{=}CH_2 \right] \ + \ AcOH$$
with OAc below each S.

$$\xrightarrow{\text{step 3}} \ \left[R\text{-}\overset{+}{S}\text{=}CH_2 \ \longleftrightarrow \ R\text{-}\overset{+}{S}\text{-}CH_2 \right]$$
with ^-OAc below each S.

$$\xrightarrow{\text{step 4}} \ R\text{-}S\text{-}CH_2OAc \tag{12}$$

We now have all the four examples of the Pummerer reactions in which each of the four steps is rate-determining, and have thus completed the whole mechanistic spectrum of the reaction. Thus, each step of the reaction will be carefully examined in the light of current mechanistic data.

II. A. Step 1

The initial step of the Pummerer reaction of the sulfoxide is acylation of sulfinyl oxygen to form the acyloxysulfonium salt. In fact, when dimethyl sulfoxide was treated with trifluoroacetic anhydride in methylene chloride at -60°C, the corresponding tri-fluoroacetoxysulfonium salt (11) was isolated in crystalline form (eq. 13). Upon warming this salt to room temperature, Swern observed the occurrence of the Pummerer reaction which gave the corresponding α-trifluoroacetoxysulfide (eq. 13).[22] Thus in the reaction with such a markedly electrophilic anhydride as trifluoroacetic anhydride, the formation of the acyloxysulfonium salt occurs beyond any doubt in the initial step of the Pummerer reaction.

$$CH_3\text{-}S\text{-}CH_3 \ + \ (CF_3CO)_2O \ \xrightarrow[CH_2Cl_2]{-60°C} \ CH_3\text{-}\overset{+}{S}\text{-}CH_3 \quad CF_3COO^-$$
with O below the first S and $O\text{-}C\text{-}CF_3$ (with O double bond) below the product S.

$$\underline{11} \tag{13}$$

$$\xrightarrow[CH_2Cl_2]{\sim 25°C} \ CH_3\text{-}S\text{-}CH_2\text{-}O\text{-}C\text{-}CF_3$$
with O double bond under the C.

Iwanami, Arita and Takahashi carried out a kinetic study on the

Pummerer reaction of dimethyl sulfoxide with p-substituted benzoic anhydrides and observed a good Hammett correlation between the rates and σ-values with a relatively large positive ρ value (+ 1.40), and also a small kinetic isotope effect (k_H/k_D= 1.21) with DMSO-d$_6$.[23] These observations suggest that even with a moderate electrophile such as benzoic anhydride, the rate-determining step of the reaction is the initial acylation.

Another example to suggest the rate-determining acylation is the intramolecular cyclization by the Pummerer reaction as shown below (eq. 14).[24-26]

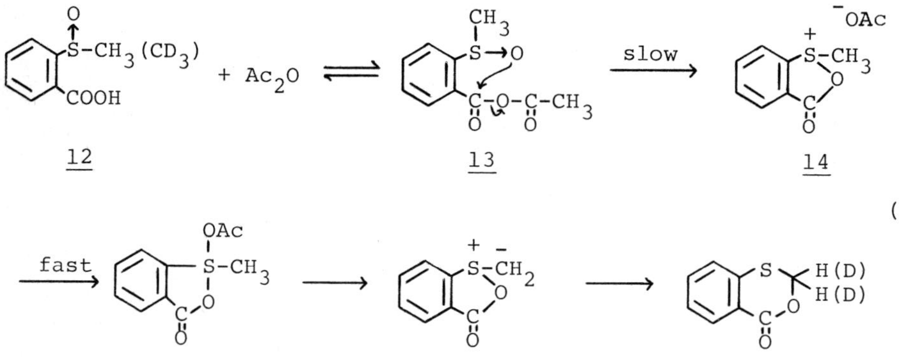

$$
\begin{array}{ccc}
\underset{\textbf{12}}{\text{(o-CH}_3\text{S-C}_6\text{H}_4\text{-COOH)}} + \text{Ac}_2\text{O} \;\rightleftharpoons\; \underset{\textbf{13}}{\text{(13)}} & \xrightarrow{\text{slow}} & \underset{\textbf{14}}{\text{(14)}}
\end{array}
$$

(14)

$$
\xrightarrow{\text{fast}} \;\underset{\textbf{14'}}{\text{(14')}}\; \longrightarrow \; \underset{\textbf{15}}{\text{(15)}}
$$

When the sulfoxide (12)-methyl-d$_3$ was allowed to react with acetic anhydride to 50% completion, the recovered sulfoxide was found to retain all three deuterium atoms completely, while the cyclized product (15) also held two deuterium atoms per molecule. Meanwhile the kinetic isotope effect was quite small, k_H/k_D = 1.07. The reactivity falls in the following order of alkyl substituents, i.e. i-Pr > n-Pr > Et > Me > CH$_2$Ph, which is the order of basicities of these alkyl phenyl sulfoxides and clearly supporting the rate-determining step to be the acylation of the sulfinyl oxygen terminus. An interesting observation is that the rate of this Pummerer cyclization reaction is 140 times larger than that of the normal Pummerer reaction of aryl methyl sulfoxide, undoubtedly due to anchimeric assistance of the neighboring ortho carboxyl group.

The initial step of this reaction is believed to be the formation of the mixed acid anhydride (13), which is considered to be a much better acylating agent, and which acylates intramolecularly the sulfoxide in the rate-determining step. The formation of the mixed acid anhydride (13) is supported by the spectroscopic observation

of a similar mixed acid anhydride (13') in the treatment of o-iso-
propylsulfinylbenzoic acid with diphenyl ketene (eq. 15).[26]

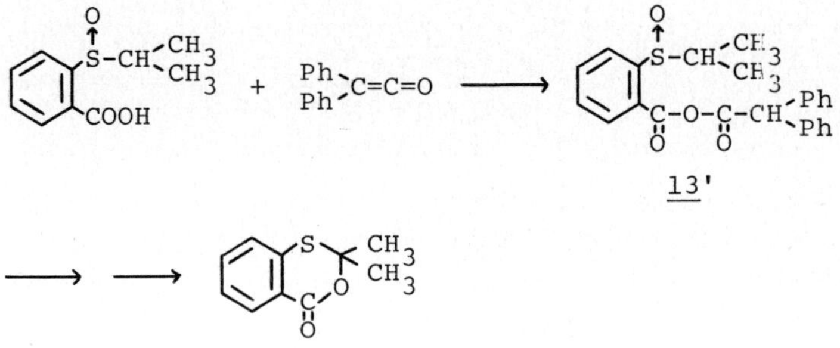

(15)

13'

Even in the cases in which other steps are rate-determining,
such as in the reaction of aryl methyl sulfoxides with acetic
anhydride, the initial acetylation has to pass through a considerably
high energy barrier in view of the relatively large negative Hammett
ρ value, -1.6.[27,28]

II. B. Step 2

The second step is the proton removal from the acyloxysulfonium
salt formed by the initial acylation of the sulfoxide to form an
acyloxysulfonium ylide (18) which was neither isolated nor even
confirmed spectroscopically. However, in view of the sizable magnitude
of kinetic isotope effect in a number of cases, the rate-determining
proton removal by acylate ion is quite evident in many Pummerer
reactions (eq. 16). We first observed a sizable kinetic isotope effect
in the reaction between aryl methyl sulfoxide and acetic anhydride,
k_H/k_D = 2.9 .[27,28] In this case the Hammett ρ value obtained by

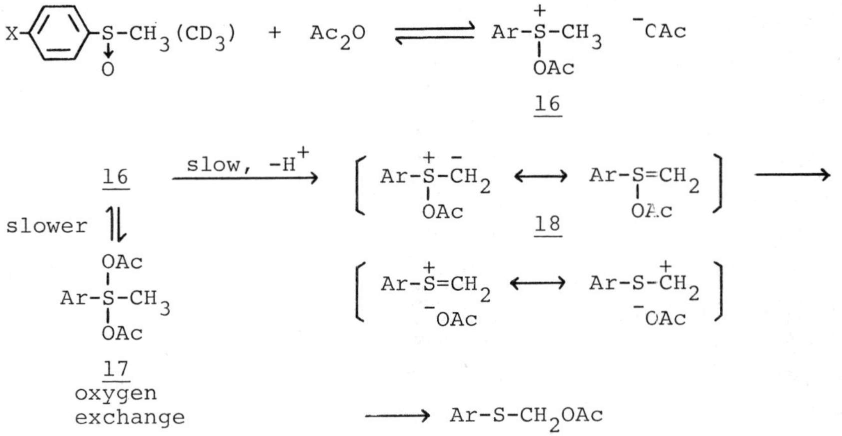

16

(16)

plotting σ-values of p-substituents was -1.6, implying that the
acylation equilibrium is also important in the energy profile of
the reaction. Meanwhile, sulfoxides are known to undergo oxygen
exchange with acetic anhydride,[29-31] presumably via a sulfurane
type intermediate (17), and this oxygen exchange is known to be
responsible for the racemization of optically active sulfoxides in
our ^{18}O tracer experiments.[29,30] Thus when an optically active
sulfoxide is subjected to the Pummerer reaction, both the Pummerer
reaction and the racemization due to oxygen exchange take place
concurrently. However, in the case of aryl methyl sulfoxide, the
Pummerer reaction proceeds about 6 times faster than the oxygen
exchange at 120°C (E_a for the Pummerer reaction is 21.2 kcal/mole,
$\Delta S^{\ddagger}$= -20.7 e.u.).[28]

   The value of the kinetic isotope effect, k_H/k_D, remained about
3 even with p-nitrophenyl methyl sulfoxide. Also a stable selenurane,
the tetracoordinated seleno-analog of sulfurane, which upon heating
underwent the Pummerer reaction, was isolated in a similar reaction of
the selenide with benzoyl peroxide (BPO).[32,33] Thus there has been
a suggestion that a sulfurane type intermediate (17) may also be
formed in the Pummerer reaction and that proton removal is taking
place intramolecularly by the acetoxy carbonyl group in the rate-
determining six-membered cyclic transition state with an angle close
to 110°-120° (eq. 17). Incidentally, the average maximum value of
the kinetic isotope effect (k_H/k_D) in the E_i reaction, which proceeds
through either 5- or 6- membered cyclic transition states, is found
to be generally in the range of 3-4.[34] However, once the sulfurane
type intermediate is formed, there would not be any asymmetric
induction at the α-position during the acetoxy migration from sulfur
to α-carbon, since the chirality around sulfur is completely lost in th
formation of the sulfurane intermediate. Moreover, oxygen exchange

$$(17)$$

17   18

should be faster than or at least equivalent to the rate of the
Pummerer reaction in this mechanistic scheme, but this was not found

to be the case.

Recently a sulfurane (19) bearing α-methylene protons was prepared by Lau and Martin and treated with NaOD-pyridine-d_5.[35] The sulfurane

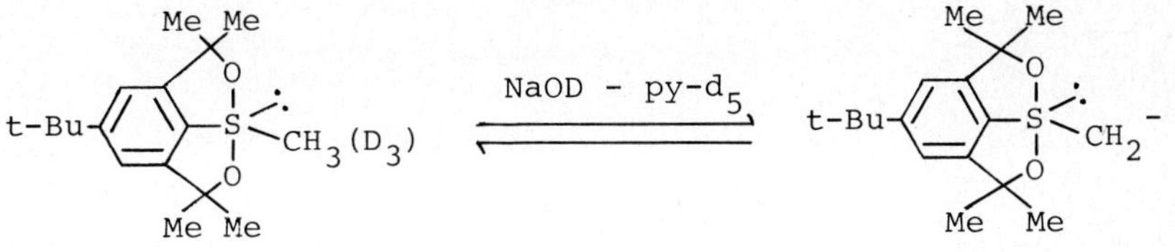

$$\text{(18)}$$

underwent a slow H-D exchange but did not give any rearrangement or decomposition product, unlike the Pummerer reaction of alkyl aryl sulfoxide. The carbanion (20) formed by proton removal from the sulfurane (19) must be quite different from the acyloxysulfonium ylide (18) formed during the Pummerer reaction, since the carbanion (20) is so rigidly built that internal bond migration would not be as facile as in the case of the sulfonium ylide (18). Moreover, the α-proton abstraction from the sulfurane intermediate is considered to be less favorable than from the acyloxysulfonium salt in view of the less acidic nature of the α-proton of the sulfurane relative to the acyloxysulfonium salt. The inductive effect of the phenyl-dialkoxysulfuranyl group (σ_I= 0.40) is found to lie between those of phenylthio group (0.21) and of the phenylsulfinyl group (0.51) based on the observations of ^{19}F NMR measurements performed with F-C_6H_4-S(X)-Ph. The value of the inductive effect for the alkoxy-sulfonium cation is very much larger (1.31), while the resonance parameters (σ_R) were obtained as -0.08 (sulfide), -0.01 (sulfoxide), 0.09 (sulfurane), 0.14 (sulfone) and 0.31 (alkoxysulfonium group).[36] Thus, the possibility of such a sulfurane intermediate in the normal Pummerer reaction of open-chain sulfoxides seems to be quite unlikely. However, in some cases, e.g., the reaction of o-carboxyphenyl alkyl sulfoxide (12) which proceeds via a cyclic acyloxysulfonium ylide to form the cyclized product (15), the intervention of a sulfurane intermediate (14') is quite conceivable (eq. 19, see also eq. 14).[26]

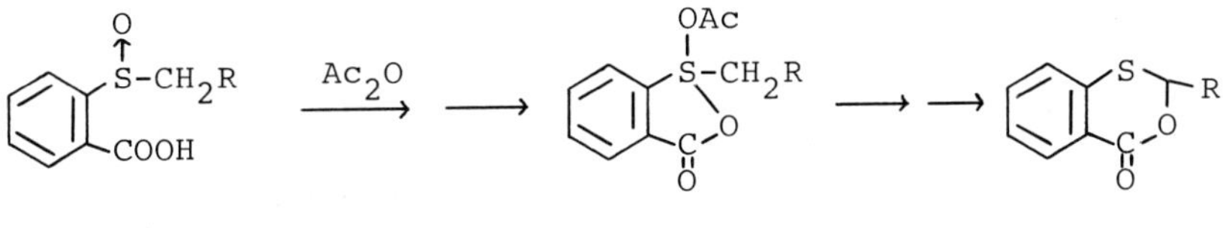

$$\text{(19)}$$

When the proton removal takes place clearly by the intermolecular path, a large kinetic isotope effect is observed. The following is one example of such a reaction.[37]

$$
\begin{array}{c}
R\overset{+}{-}S\text{-}CH_3(CD_3) \\
O=C\diagdown N \diagup C=O \quad Cl^- \\
\end{array}
\quad + \quad R_3N \quad \xrightarrow{<0°} \quad
\begin{array}{l}
R\text{-}S\text{-}CH_2(D_2)\overset{+}{-}NR_3 \\
\text{or} \quad R\text{-}S\text{-}CH_2(D_2)\text{-}N\diagup CO \diagdown CO
\end{array}
\qquad (20)
$$

$$k_H/k_D = 9\text{-}10$$

Wilson and Strong also obtained a large kinetic isotope effect (k_H/k_D) of about 9 in the competitive Pummerer reaction of α,α-dibenzyl sulfoxide with acetic anhydride.[38] This value of the kinetic isotope effect seems somewhat too high, but indicates clearly that proton removal is the rate-determining step of the reaction.

$$
\begin{array}{c}
Ph\text{-}CH_2\text{-}\underset{O}{\overset{\downarrow}{S}}\text{-}CD_2\text{-}Ph \quad + \quad Ac_2O \quad \longrightarrow \quad
\underset{OAc}{Ph\text{-}CH\text{-}S\text{-}CD_2\text{-}Ph} \\
\\
+ \quad \underset{OAc}{Ph\text{-}CH_2\text{-}S\text{-}CD\text{-}Ph}
\end{array}
\qquad (21)
$$

A large primary deuterium isotope effect of at least 4 was also observed in the reaction of dimethyl sulfoxide (or DMSO-d_6) with acetyl chloride.[39]

$$
CH_3\text{-}\underset{O}{\overset{\downarrow}{S}}\text{-}CH_3 \; + \; CH_3COCl \; \longrightarrow \;
\underset{OCOCH_3}{CH_3\text{-}\overset{+}{S}\text{-}CH_3} \; Cl^- \; \longrightarrow \;
\underset{Cl}{CH_3\text{-}\overset{+}{S}\text{-}CH_3} \; {}^-OAc
\qquad (22)
$$

$$\longrightarrow \quad CH_3\text{-}S\text{-}CH_2Cl$$

II. C. Regioselectivity

The importance of proton removal may be found in the regioselectivit of the Pummerer reaction. In other words, since the acyloxy migration takes place from the sulfur to the α-carbon atom from which the proton is removed, the migratory aptitude is determined by the relative acidity of α-hydrogen atoms in the acyloxysulfonium salt. The more stable acyloxysulfonium ylide is formed preferentially rather than the less stable one, upon removal of an α-proton prior to the acyloxy migration, in keeping with the mechanism involving the acyloxysulfonium ylide intermediate. Thus one can see a clear-cut regioselectivity in the Pummerer reaction which is useful for many organic syntheses.

In the reaction of alkyl methyl sulfoxides with acetic anhydride, the acetoxy group migrates only to the methyl moiety (eq. 23),[8]

$$RCH_2-S-CH_3 \ + \ Ac_2O \ \longrightarrow \ RCH_2-S-CH_2OAC \qquad\qquad (23)$$

$$RCH_2 \ : \ n-Pr, \ i-Pr, \ n-Bu$$

while with p-nitrobenzyl benzyl sulfoxide, acetoxy migration occurs only to the p-nitro-benzylic site selectively (eq. 24).[40] A more

$$p-NO_2C_6H_4CH_2-S-CH_2Ph \ \xrightarrow{AcOH} \ \left[\ p-NO_2C_6H_4\underset{OAc}{CH}-S-CH_2Ph \ \right]$$

$$\xrightarrow{H^+ \ or \ \Delta} \ 1/2 \ p-NO_2C_6H_4CH(SCH_2Ph)_2 \qquad\qquad (24)$$

clear-cut case is the Pummerer reaction of cyanomethyl benzyl sulfoxide in which the acetoxy group migrates only to the cyanomethyl side (eq. 25).[41]

$$PhCH_2-S-CH_2CN \ + \ Ac_2O \ \longrightarrow \ PhCH_2-S-\underset{OAc}{CHCN} \qquad\qquad (25)$$

In the earlier study on the Pummerer reaction of methionine sulfoxide, migration of the acetoxy group was found to take place nearly exclusively to the methyl side, thus giving homocysteine upon hydrolysis.[1] Thus migration of the acetoxy group generally takes place at α-carbon which has the most acidic proton. Qualitatively the acidity of α-methylene proton of various substituted sulfoxides falls in the following order, as does also seem to be the regiospecificity:

$$CH_2\underset{O}{CR}, \ CH_2CN > PhCH_2 > CH_3 > n-alkyl > sec.-alkyl$$

One intriguing observation is that in the reaction of a methyl-sulfinyl sugar (21) with acetic anhydride, acetoxy migration occurred preferentially at the methylenic site rather than the methyl site to the extent of 2 : 1 (eq. 26).[42]

No detailed information on the conformational arrangement of the sugar is known; however, some of the acetoxy groups and also the etheral oxygen atom might be participating in the reaction resulting in preferential proton removal from the methylenic site.

$$
\underset{\underline{21}}{\text{(structure)}} \quad \xrightarrow[100^\circ C]{Ac_2O} \quad \underset{2}{\text{(structure)}} \quad + \quad \underset{1}{\text{(structure)}} \tag{26}
$$

CH$_2$-S-CH$_3$ (with O↑ and O), R, R, R, OMe, R **21**

R-CH-S-CH$_3$ (with O), R, R, OMe, R **2**

CH$_2$-S-CH$_2$R (with O), R, R, OMe, R **1**

R = OAc

Earlier, the acetoxy group was reported by Parham et al. to migrate preferentially to the benzylic site in the reaction of benzyl methyl sulfoxide with acetic anhydride.[43] However, our recent reinvestigation revealed that the migration occurs both to the benzylic and the methyl carbons in the ratio of 2 : 1 (eq. 27).[44]

$$
PhCH_2\text{-}\underset{O}{\overset{\downarrow}{S}}\text{-}CH_3 \; + \; Ac_2O \; \xrightarrow{80\text{-}100^\circ C} \; 1/2 \; PhCH(SCH_3)_2 \; + \; PhCHO
$$

50-70%

$$ \tag{27} $$

+ PhCH$_2$-S-CH$_2$OAc

30-34%

The benzylic protons are more acidic than the methyl protons and would more readily be removed by acetate ion, but not predominantly, and this is due mainly to the steric shielding by the phenyl group. Incidentally, the Pummerer reactions of dibenzyl and benzyl phenyl sulfoxides are not facile and require relatively higher temperatures than would be expected from the acidity of benzylic protons.[38,45,46] The relatively low basicity of the benzylic sulfoxide oxygen terminus is also partly responsible for the somewhat sluggish reactivity of the Pummerer reaction.

A good example illustrating the change of regioselectivity in the Pummerer reaction due to steric strain was reported by Jones et al. as shown in eq. 28.[47] The α-isomeric steroid bearing a methylsulfinyl group at the 6-position (22) afforded the 6-acetoxy-methylthio derivative, the normal Pummerer product. Meanwhile, with the β-isomer (23) the corresponding acetoxysulfonium salt (24) is sterically crowded due to the two axial hydrogen atoms at the 4- and 8- positions and the one axial methyl group (19-CH$_3$); thus proton removal from the α-ring carbon would result in the release of steric repulsion forming a nearly coplanar ylide (25). Thus, the acetoxy migration is presumed to take place at the α-ring carbon, followed by subsequent elimination of acetic acid to result in the olefinic steroid (26).

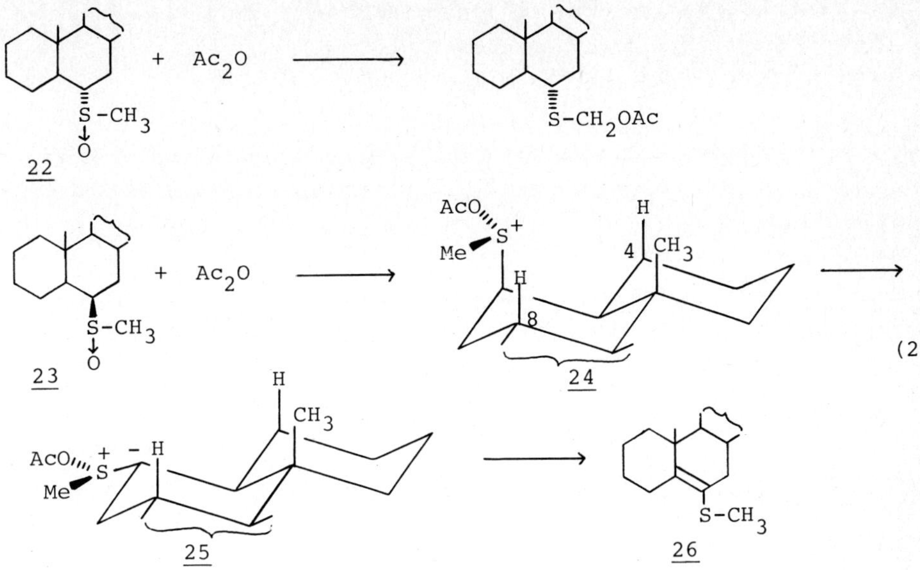

(28)

II. D. Step 3

The third step of the Pummerer reaction is considered to be the
concurrent electron transfer within the acyloxysulfonium ylide and
S-O bond cleavage, which results in the formation of an ion pair
between the sulfur-stabilized carbonium ion and the acetate ion.
The sulfur-stabilized cation may or may not be an intimate ion pair,
depending upon the nature of the substituent R' (eq. 29), while the
life time of the ion pair also depends very much on the substituent
R' and other environments, such as solvent, acylating agent, etc.
Unfortunately none of these intermediates have been detected.
However, there are reactions in which the third step is obviously
rate-determining and we call these "ElcB" Pummerer reactions. In this
reaction proton removal is fast and reversible, and S-O bond cleavage
is the slowest step (eq. 29). Usually the R' group is electron-withdrawing such as cyano, carbonyl, acetylenic or cyclopropyl, so as to
facilitate the proton removal, while S-O bond cleavage is retarded
by the strong inductive effect of R' group, as in the ElcB mechanism

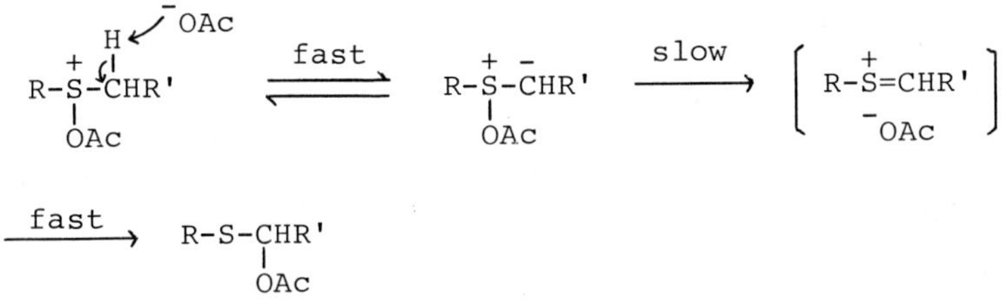

(29)

for olefin formation. The following four sulfoxides are known to proceed via this ElcB type mechanistic route:

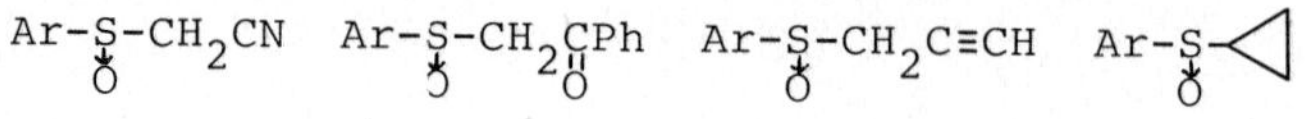

$$Ar-\underset{O}{\overset{*}{S}}-CH_2CN \quad Ar-\underset{S}{\overset{*}{S}}-CH_2\underset{O}{\overset{O}{\overset{\|}{C}}}Ph \quad Ar-\underset{O}{\overset{*}{S}}-CH_2C\equiv CH \quad Ar-\underset{O}{\overset{*}{S}}{-}\triangleleft$$

In the typical case of cyanomethyl phenyl sulfoxide, practically no kinetic isotope effect (k_H/k_D = 1.02) was observed in the reaction with acetic anhydride,[15] while the cyanomethyl-d_2 derivative was found to lose deuterium completely before half-completion of the reaction in a fast reversible proton removal pre-equilibrium. The rate-determining step is believed to be S-O bond cleavage, similar to the ElcB type of elimination.

The ElcB type mechanism is more evident in the Pummerer type reactions of sulfimides and sulfonium ylides. A well studied case is the reaction between the 5-membered N-tosylsulfimide (27) and alcoholic potassium hydroxide to give the corresponding α-alkoxy-sulfide shown below (eq. 30).[48] The kinetic isotope effect (k_H/k_D = 1.09) has the size of a secondary effect due to hyperconjugation, which one often finds in solvolysis, while the recovered sulfimide was found to have lost its original deuterium during the reaction. Meanwhile, the Hammett ρ value obtained from the satisfactory correlation of the rates with σ-values was markedly large, i.e., + 2.0. All these observations fit the ElcB type of mechanism. Here

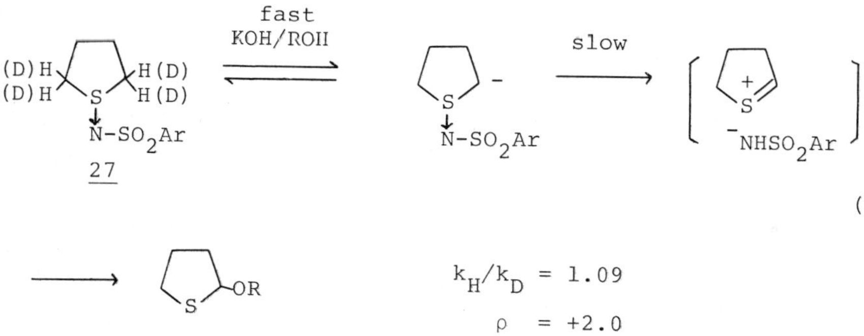

$$k_H/k_D = 1.09$$
$$\rho = +2.0$$

Ar : p-CH$_3$C$_6$H$_4$, C$_6$H$_5$, p-ClC$_6$H$_4$

again, the original sulfimide has a poor leaving group, namely the arenesulfonamide ion. In the case of aryl alkyl N-tosylsulfimides, treatment with alcoholic alkali hydroxide gives only a portion of the Pummerer type reaction product and mainly the reduced sulfides, presumably formed by initial nucleophilic attack of alkoxide on the sulfur atom as shown below (eq. 31).[49] When alkyl N-tosylsulfimides

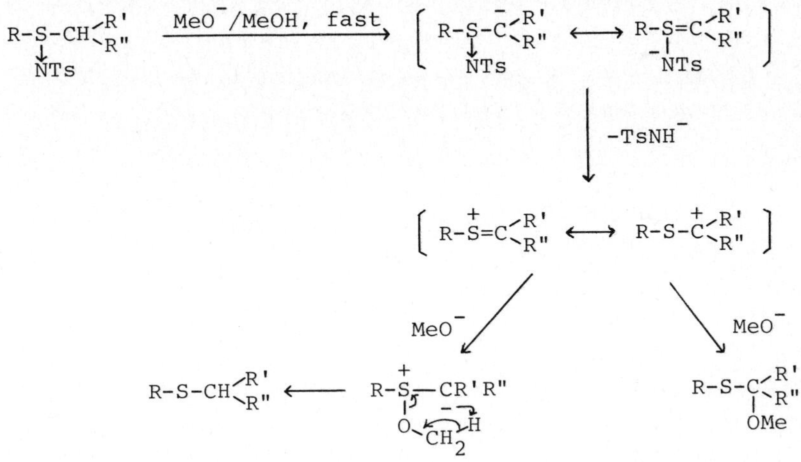

$$R-S-CH\overset{R'}{\underset{R''}{<}} \quad \xrightarrow{MeO^-/MeOH, \; fast} \quad \left[R-\overset{-}{\underset{NTs}{S}}-C\overset{R'}{\underset{R''}{<}} \longleftrightarrow R-\overset{+}{\underset{NTs}{S}}=C\overset{R'}{\underset{R''}{<}} \right]$$

$$\Big\downarrow -TsNH^-$$

(31)

$$\left[R-\overset{+}{S}=C\overset{R'}{\underset{R''}{<}} \longleftrightarrow R-S-\overset{+}{C}\overset{R'}{\underset{R''}{<}} \right]$$

MeO$^-$ / \ MeO$^-$

$$R-S-CH\overset{R'}{\underset{R''}{<}} \longleftarrow R-\overset{+}{\underset{O-CH_2}{S}}-CR'R'' \qquad R-S-\overset{R'}{\underset{OMe}{C}}\overset{R'}{\underset{R''}{<}}$$

having β-hydrogens were treated with t-BuOK in benzene, the corresponding vinyl sulfides were obtained in over 50% yield.[50]

Aryl methyl N-p-tosylsulfimides undergo the Pummerer reaction
with acetic anhydride with nearly equal ease. However, the values
of the activation parameters ($\Delta H^{\ddagger}$= 15.2 kcal/mole, $\Delta S^{\ddagger}$= -41.2 e.u.)[51]
are quite similar to those of the S_N2 type oxygen exchange reactions
of diaryl sulfoxides with acetic anhydride ($\Delta H^{\ddagger}$= 13.1 kcal/mole,
$\Delta S^{\ddagger}$= -45.1 e.u.).[29,30] Moreover, the corresponding sulfoxides are
isolated when the reaction is stopped at half completion. These
observations fit the following mechanism which involves the rate-
determining S_N2 acetoxy exchange shown below (eq. 32). The values
of the Hammett ρ values (ρ_X= -0.71, ρ_Y= -0.57) and the small kinetic
isotope effect (k_H/k_D = 1.57) seem to support this mechanism.[51]

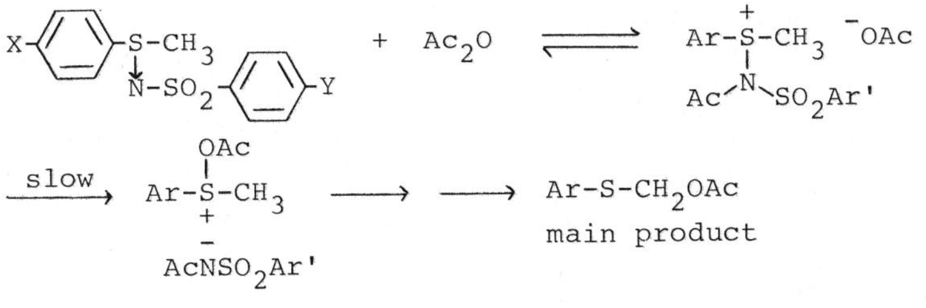

$$X-\underset{N-SO_2}{\overset{S-CH_3}{\bigcirc}}-\bigcirc-Y \quad + \; Ac_2O \; \rightleftharpoons \; \overset{+}{Ar-\underset{Ac^N\diagdown SO_2Ar'}{S}-CH_3} \quad {}^-OAc$$

(32)

$$\xrightarrow{slow} \quad Ar-\overset{OAc}{\underset{+}{S}}-CH_3 \quad \longrightarrow \longrightarrow \; Ar-S-CH_2OAc$$
$$\underset{AcNSO_2Ar'}{} \qquad\qquad\qquad\qquad \text{main product}$$

Treatment of a stable sulfonium ylide, such as (28), with either
acetic anhydride or benzoyl peroxide gives similarly the Pummerer
reaction product, namely α-acetoxymethyl aryl sulfide or α-benzoyloxy-
methyl aryl sulfide in quantitative yields (eq. 33).[52] The kinetic

isotope effect (k_H/k_D) observed with the trideuterated methyl aryl
sulfonium ylide (28) and acetic anhydride was 1.57 while that with
BPO was as small as 1.13. Meanwhile, the ylide (28), recovered after

$$X-\langle C_6H_4 \rangle-\overset{+}{\underset{\underset{C(COOMe)_2}{|}}{S}}-CH_3 \quad + \quad Ac_2O \quad \longrightarrow \quad Ar-S-CH_2OAc \quad + \quad AcCH(COOMe)_2$$

28

$$\Delta H^{\ddagger} = 21.4 \text{ kcal/mole}$$
$$\Delta S^{\ddagger} = -22.2 \text{ e.u.}$$

(33)

$$+ \quad BPO \quad \longrightarrow \quad Ar-S-CH_2OCOPh \quad + \quad PhCO_2CH(COOMe)_2$$

$$\Delta H^{\ddagger} = 17.3 \text{ kcal/mole}$$
$$\Delta S^{\ddagger} = -18.7 \text{ e.u.}$$

half completion of the reaction, was found to have lost over 80% of
the original deuterium label, revealing the reaction to proceed via
a typical ElcB mechanistic path. The activation parameters for the
reaction are of the same order of magnitude as those of the normal
Pummerer reaction of the corresponding sulfoxide ($\Delta H^{\ddagger} = 21.2$ kcal/mole,
$\Delta S^{\ddagger} = -20.7$ e.u.).[27,28]

II. E.   Step 4 and the Nature of the Ion Pair

Step 4 involves recombination of the acyloxy group with the
sulfur-stabilized carbonium ion within the ion pair which is formed
by the S-O bond cleavage in Step 3. The shift is believed to be
quite fast, since it involves the recombination of two oppositely
charged ion species either within the ion pair or in a close vicinity.
The final product is the rearranged ester formed in this last step;
however, the mode of the acetoxy shift differs from one sulfoxide
to another and depends also on the nature of the migrating group.
Therefore, there has been some controversy on the nature of the
migration.

It was believed that the ion pair is formed through a typical
E2 type concerted elimination process involving simultaneous proton
removal and cleavage of acetate group from the acetoxysulfonium salt.
The transition state of the reaction is then represented by the
formula (29) in which a positive charge would be developed at the

$$R-\underset{\underset{OAc}{|}}{S}-CH_2R' \quad \longrightarrow \quad \left[\underset{\underset{\delta-OAc}{|}}{\overset{\overset{\overset{H---base}{|}}{\delta+}}{R-S=\!=\!=CHR'}} \right]^{\ddagger} \quad \longrightarrow \quad \left[R-\overset{+}{S}=CHR' \quad {}^-OAc \right]$$

29

α-carbon as the cleavage of the acetoxy group proceeds. In this case the transition state must be stabilized more by an electron-donating branched alkyl group than by an unbranched alkyl group such as methyl, which is contrary to observation; namely, the preferential acetoxy migration to the less branched α-alkyl carbon site. Thus the possibility of the E2 type elimination is ruled out.

The most controversial point of discussion has been whether the migration is intramolecular or intermolecular. Following our proposal of the intermolecular mechanism for acetoxy migration on the basis of the ^{18}O-tracer investigation,[7] the intermolecular mechanism prevailed for more than a decade. However, recent stereochemical investigations have revealed the occurrence of the intramolecular acetoxy migration in various cases.[14,15] Therefore let us consider the intramolecular stereospecific Pummerer reactions in the following section.

II. F. Stereospecific Intramolecular Pummerer Reactions

When optically active cyanomethyl p-tolyl sulfoxide (30) was treated with acetic anhydride at 120°C for 3.5 hours, we obtained in a good yield the corresponding α-acetoxy-α-cyanomethyl sulfide (31) which was found to have an induced optical rotation at the α-carbon to the extent of 29% e.e. upon measurement with europium shift reagent (eq. 34).[15]

$$CH_3 \text{—⟨⟩—} S\text{-}CH_2CN + Ac_2O \xrightarrow{120°} CH_3\text{—⟨⟩—} \underset{\underset{OAc}{|}}{\overset{\overset{H}{|}}{S\text{-}C\text{-}CN}} \qquad (34)$$

$[\alpha]_D$ +252° $[\alpha]_D$ +26.8°
                                (e.e., 29%)

30   31

The original ^{18}O-labeled sulfoxide, recovered after 50% completion of the reaction, was found to retain 96% of the original label while the product ester (31) was found to contain 85% of [18]O of the original sulfoxide. When the loss of 4% [18]O due to the possible oxygen exchange is taken into account, the reaction is considered to proceed via the intramolecular acetoxy migration at least to the extent of 90%. Meanwhile, a careful analysis of the distribution of [18]O in the ester revealed that 63% of [18]O is located at the carbonyl group while the remaining 37% is at the etheral oxygen. This uneven distribution of [18]O alone can suggest that the precursor of the ester is not a dissociated ion pair but is closely bound as an intimate ion pair or even an undissociated ylide-like intermediate which is still covalently bound.

The rate-determining step of this reaction is considered to be the cleavage of the S-O bond of the acetoxysulfonium ylide (32) in the ElcB type process, while the effect of p-substituent on the rate of the reaction may be represented by the Hammett ρ value of -0.65. The retention of 85% of ^{18}O of the original sulfoxide into the resulting ester and the substantial asymmetric induction at the α-carbon suggest that the acyloxy migration occurred intramolecularly via an intimate ion pair. The uneven distribution of ^{18}O, i.e., more in carbonyl than in ether oxygen, reveals that step 4, namely the recombination, is very fast and suggests that the migration proceeded via both a 5-membered cyclic and a 3-membered sliding mode concurrently at the transition state in which the former is the major route, due probably to the necessary anchimeric assistance of the carbonyl group to ease the S-O bond cleavage (eq. 35).[15]

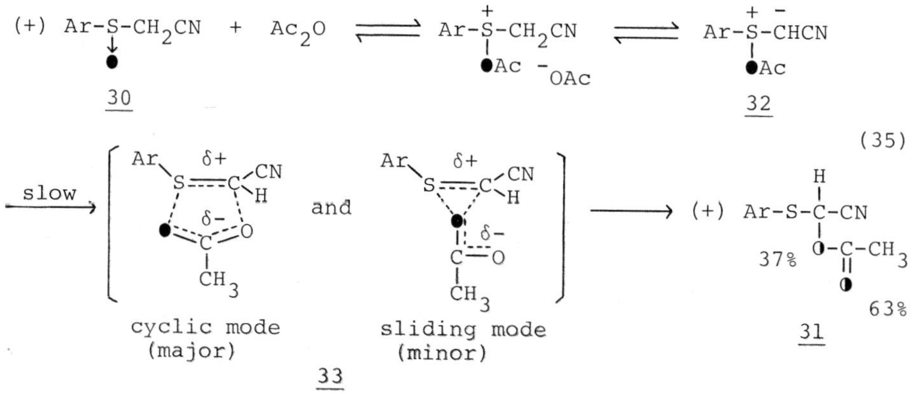

$$(35)$$

The asymmetric induction cannot be caused by the stereoselective proton removal from the acyloxysulfonium salt, since the resulting α-cyanocarbanion, the ylide (32), is known to be a coplanar resonance stabilized sp^2 carbanion and the proton removal is a reversible process. Therefore, the asymmetric induction should have taken place just when the acetoxy group is shifted from the chiral sulfur atom to the α-carbon. One might be intrigued by the dual pathways for the acetoxy migration in which the five-membered cyclic path predominates over the three-membered sliding path. However, similar dual paths for acyloxy migration have been known in several rearrangements of t-amine oxides with acylating agents.[53] Meanwhile, there have been many examples of similar acyloxy migrations via intimate ion pairs also in the rearrangements of heteroaromatic N-oxides with acylating agents.[53]

Other examples of asymmetric induction in the Pummerer reaction exist. p-Tolyl propargyl sulfoxide (34) gives the corresponding asymmetrically induced optically active α-acetoxysulfide,[54] and so does p-tolyl phenacyl sulfoxide (35)[44] as shown below (eq. 36).

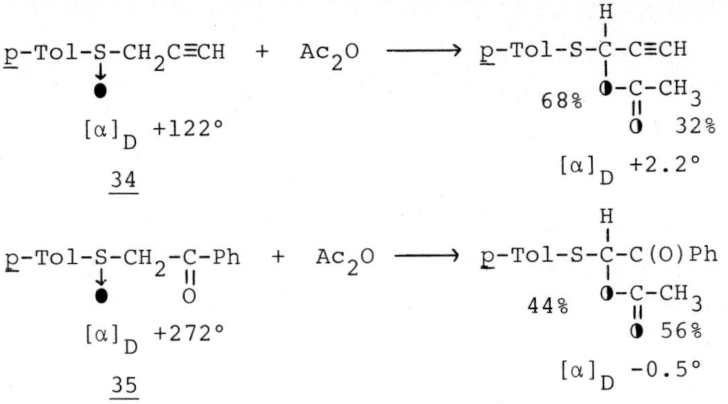

$$\text{p-Tol-S-CH}_2\text{C}\equiv\text{CH} \quad + \quad \text{Ac}_2\text{O} \quad \longrightarrow \quad \begin{array}{c} \text{H} \\ | \\ \text{p-Tol-S-C-C}\equiv\text{CH} \\ | \\ \text{O-C-CH}_3 \\ || \\ \text{O} \end{array}$$

$[\alpha]_D$ +122°

68%   32%

34

$[\alpha]_D$ +2.2°

(36)

$$\text{p-Tol-S-CH}_2\text{-C-Ph} \quad + \quad \text{Ac}_2\text{O} \quad \longrightarrow \quad \begin{array}{c} \text{H} \\ | \\ \text{p-Tol-S-C-C(O)Ph} \\ | \\ \text{O-C-CH}_3 \\ || \\ \text{O} \end{array}$$

$[\alpha]_D$ +272°

44%   56%

35

$[\alpha]_D$ -0.5°

In both cases, the usual ^{18}O tracer experiment with ^{18}O-labeled sulfoxides revealed that acetoxy migration has proceeded intra-molecularly at least to the extent of more than 50%, while the distribution of ^{18}O was found to be uneven, suggesting strongly that the reactions are intramolecular processes which proceed via diastereomeric intimate ion pairs. One obvious case of an intra-molecular Pummerer reaction to give asymmetrically induced product is the reaction of (+)-o-benzylsulfinylbenzoic acid with acetic anhydride found by Allenmark et al. (eq. 37).[14] In this case the

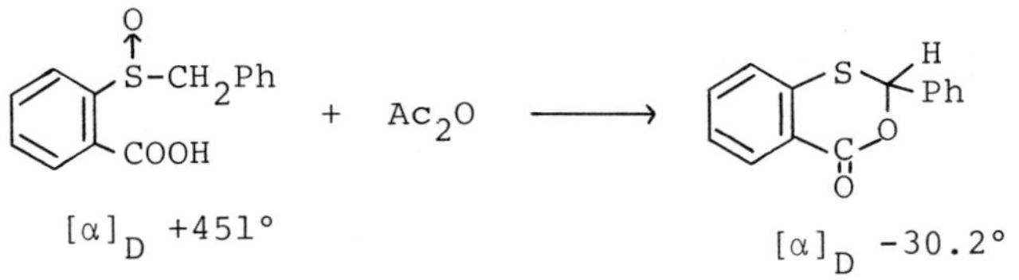

$[\alpha]_D$ +451° $[\alpha]_D$ -30.2°   (37)

extent of asymmetric induction was 19.5%(e.e.) at the α-carbon. When the sulfoxide was treated with dicyclohexylcarbodiimide (DCC) in THF, the resulting cyclic ester was asymmetrically induced at the α-carbon to the extent of 29.8%.[14]

Another example of asymmetric induction in the Pummerer reaction is that of α-phosphoryl sulfoxide (36) with acetic anhydride shown below (eq. 38).[55]

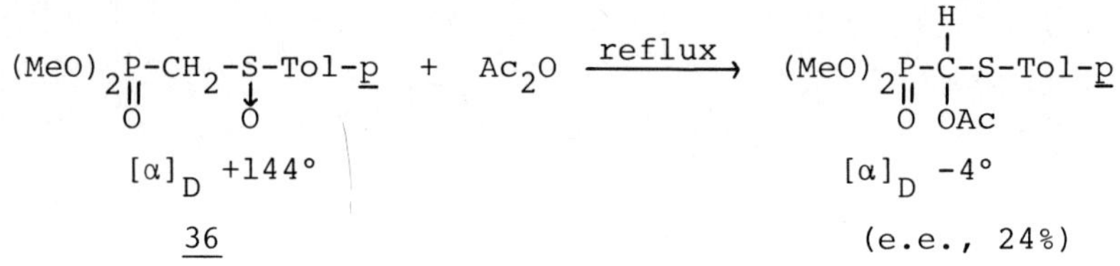

$$(\text{MeO})_2\overset{\text{O}}{\underset{\|}{P}}-CH_2-\overset{\downarrow}{\underset{\|}{S}}-\text{Tol-}\underline{p} + Ac_2O \xrightarrow{\text{reflux}} (\text{MeO})_2\overset{H}{\underset{\overset{\|}{O}}{P}}-\overset{H}{\underset{OAc}{C}}-S-\text{Tol-}\underline{p} \qquad (38)$$

$$[\alpha]_D +144° \qquad\qquad\qquad\qquad [\alpha]_D -4°$$

$$\underline{36} \qquad\qquad\qquad\qquad (e.e., 24\%)$$

In all these cases in which asymmetric induction was found to take place, the intermediate acyloxysulfonium ylides are substantially stabilized by either electron-withdrawing substituents or conjugative resonance with substituents, or both, and the heterolytic cleavage of the S-O bond generally gives highly unstable, very short-lived carbonium ions which undergo rapid recombination with the intimate anion.

Apart from the Pummerer reaction which results in asymmetrically induced products, there are a few other rearrangements involving optically active trivalent sulfur compounds which lead to asymmetric induction at the α, β, or γ carbon atom by an intramolecular rearrangement of alkyl (aryl) group from the sulfur atom to the carbanionic carbon atom of the intermediary sulfonium ylides. One is the 2,3-sigmatropic rearrangement of an allyl group of a sulfonium salt (eq. 39),[56] and the second example is the 1,4-sigmatropic rearrangement of the 10-aryl group of the 10-aryl-10-thioxanthenium salt (eq. 40),[57] while the others are the Sommelet-Hauser rearrangement of a sulfonium salt (eq. 41),[58] and that of a sulfimide (eq. 42)[59] as shown below.

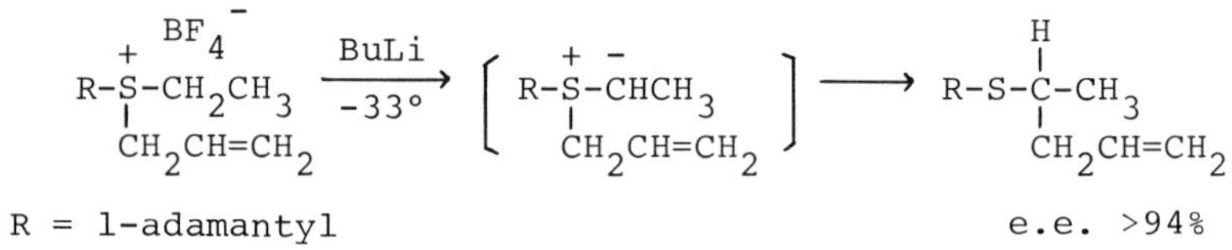

$$\underset{CH_2CH=CH_2}{\overset{+}{R-S}-CH_2CH_3} \; BF_4^- \xrightarrow[-33°]{\text{BuLi}} \left[\underset{CH_2CH=CH_2}{\overset{+}{R-S}-\overset{-}{CHCH_3}} \right] \longrightarrow \underset{CH_2CH=CH_2}{\overset{H}{\underset{}{R-S-\overset{|}{C}-CH_3}}} \qquad (39)$$

R = 1-adamantyl 　　　　　　　　　　　　　e.e. >94%

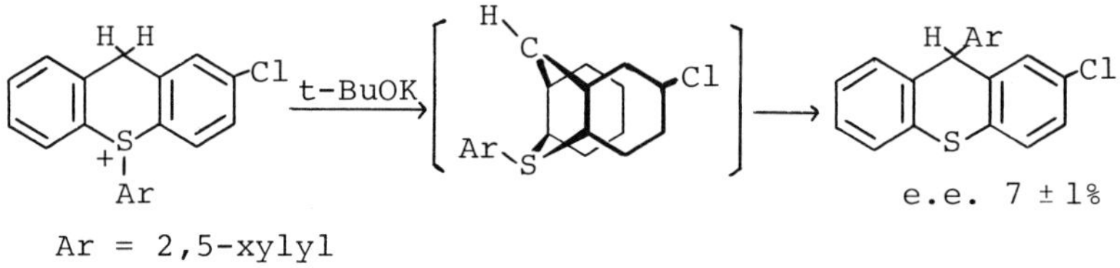

$$(40)$$

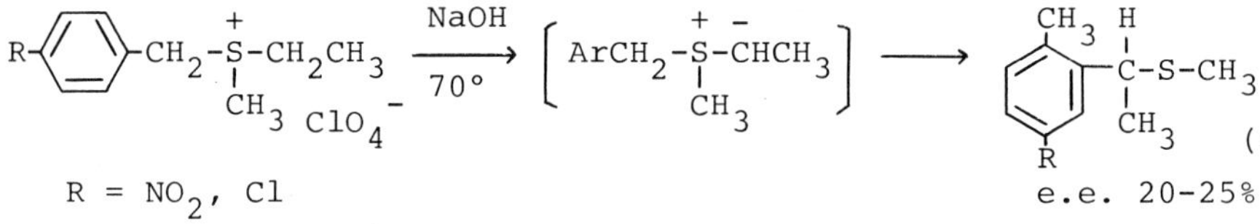

Ar = 2,5-xylyl 　　　　　　　　　　　　　e.e. 7 ± 1%

$$R-\!\!\left\langle\;\right\rangle\!\!-\underset{CH_3}{\overset{+}{CH_2-S}-CH_2CH_3} \; ClO_4^- \xrightarrow[70°]{\text{NaOH}} \left[\underset{CH_3}{ArCH_2-\overset{+}{S}-\overset{-}{CHCH_3}} \right] \longrightarrow (41)$$

R = NO₂, Cl 　　　　　　　　　　　　　e.e. 20-25%

(42)

$[\alpha]_D$ +10.7° $[\alpha]_D$ -3.3°

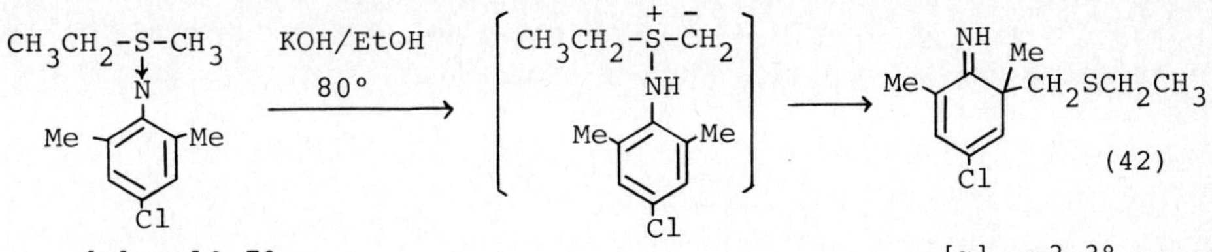

There are several other examples of stereospecific Pummerer reactions leading to the formation of stereospecific geometric isomers. One example is the Pummerer reaction of phenyl cyclopropyl sulfoxide with acetic anhydride which proceeds with 69-76% stereospecificity giving the final products in which one isomer predominates over the other (eqs. 43 and 44).[60]

76 : 24 (43)

69 : 31 (44)

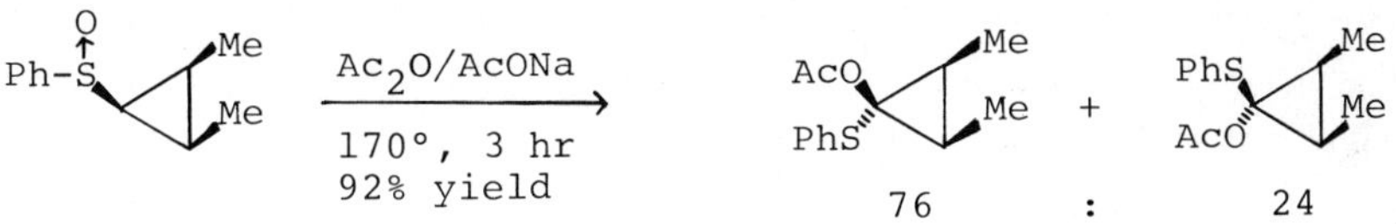

The kinetic isotope effect in this reaction found with the α-deuterated derivative (37-d) was very small, i.e., 1.13-1.49, while the Hammett plot of the rates with σ values gave a U-shape curve. The values of the activation parameters in this reaction are usually large ($\Delta H^{\ddagger}$= 41.3 kcal/mole, $\Delta S^{\ddagger}$= +10.4 e.u.) and the original D-label in the starting sulfoxide (37) is gradually lost during the reaction (eq. 45).[61] However, there is no formation of any ring opening product, while in the acetolysis of 1-chloro-1-

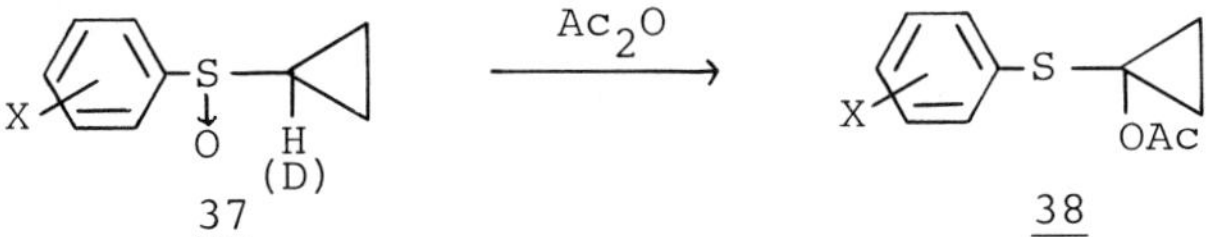

(45)

37 38

k_H/k_D ; 1.13 for X = p-MeO
1.24 for X = H
1.49 for X = m-CF_3

$\Delta H^{\ddagger}$= 41.3 kcal/mole
$\Delta S^{\ddagger}$= +10.4 e.u. (170°C)
(for X = H)

phenylthiocyclopropane which proceeds through a truely sulfur-stabilized carbonium ion intermediate, the ring opening products are formed substantially (eq. 46).[61,62]

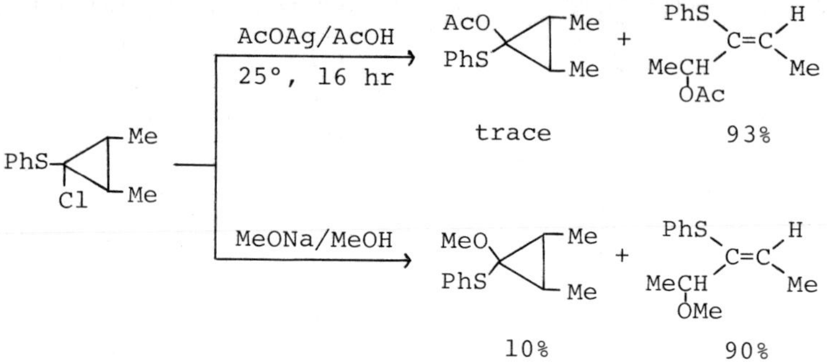

$$ (46) $$

All these observations suggest that the Pummerer reaction of these cyclopropyl sulfoxides also proceeds via an ElcB type of pathway with S-O bond cleavage at the rate-determining step. The usual ^{18}O-tracer experiment with the ^{18}O-labeled sulfoxide (37) revealed that the rearranged acetoxy derivatives (38) retained 22-30% of the original [18]O label of the sulfoxide (37), while the acetoxy group in the product is bound to the α-carbon atom opposite from the α-hydrogen which was removed. Meanwhile, the reaction is markedly facilitated by the presence of a small amount of acetic acid. These observations suggest that the transition state of the rearrangement lies close to the product and may be illustrated by the structure (39).

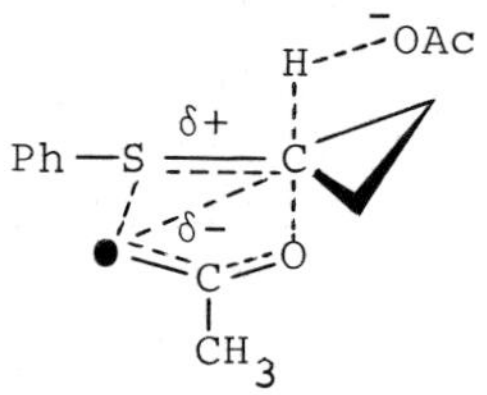

This is the first example of a Pummerer reaction in which the configuration around the α-carbon has been found to be inverted.

In connection to the sulfur-stabilized carbonium ion from 1-halocyclopropyl sulfide, there is an interesting study which indicates that there is no ring rupture when 1-halocyclopropyl sulfides were treated with alkanethiolates in basic methanol as shown below (eq. 47).[63] With CD_3SNa, the following reductive removal of halogen by C-D rupture was observed (eq. 48), clearly suggesting that the reaction follows the following path which is

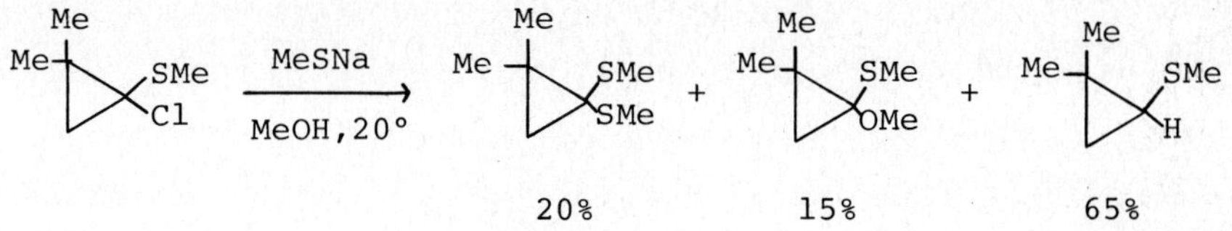

$$\text{(structures for eq. 47)} \tag{47}$$

20%   15%   65%

$$\text{(structures for eq. 48)} \tag{48}$$

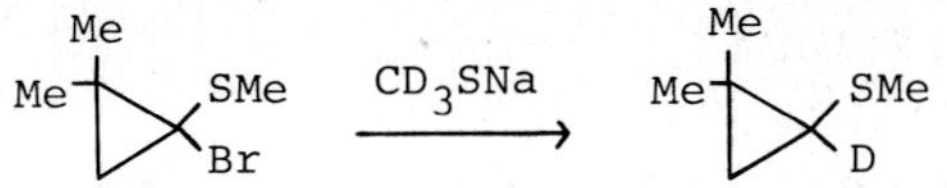

essentially the same as the Pummerer mechanism (eq. 49).

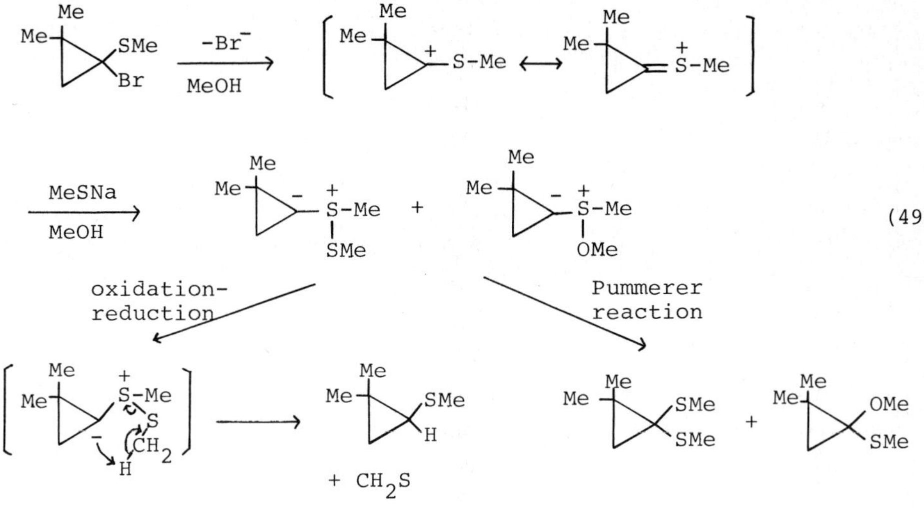

$$\text{(structures and schemes for eq. 49)} \tag{49}$$

oxidation-reduction

Pummerer reaction

+ CH$_2$S

Meanwhile, the stereochemistry of the α-halogenation of various sulfoxides seems to follow nearly the same mechanistic route, to give eventually the α-halosulfoxides in which both the configuration of the α-carbon and that of sulfinyl sulfur were found to be inverted (eqs. 50 and 51).[64,65] This somewhat complicated stereo-chemistry of α-halogenation has been accepted; however, in the

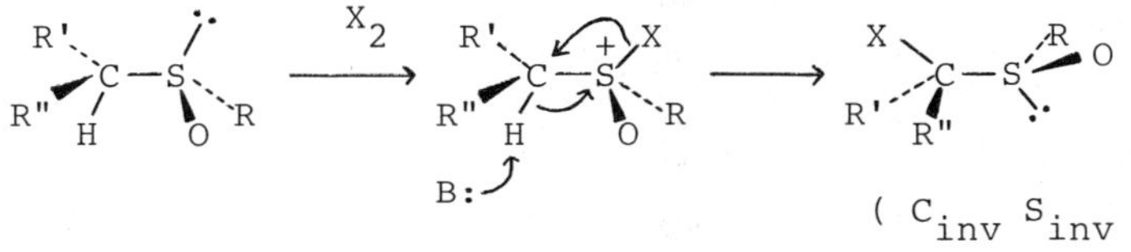

$$\text{(structures for eq. 50)} \tag{50}$$

(C_{inv} S_{inv})

Pummerer reaction, more examples of stereochemistry with either conformationally fixed sulfoxides or sulfoxides with a chiral

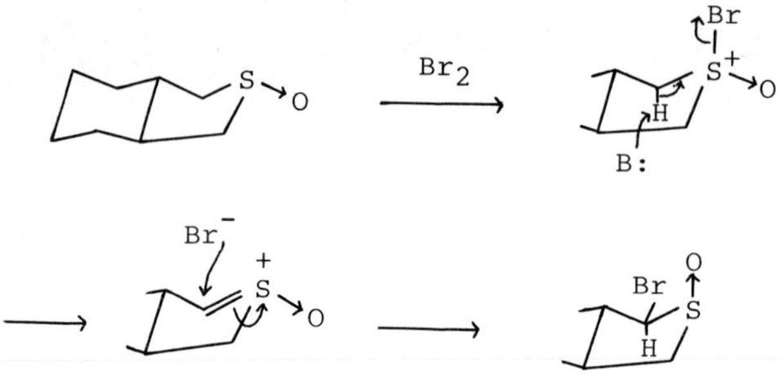

(51)

center at the α-carbon are necessary to determine the general stereochemical pattern around the α-carbon.

One of the earliest examples of a stereospecific Pummerer reaction is the reaction of a five-membered cyclic sulfoxide, i.e. 2,2-di-alkyl-1,3-oxathiolan-5-one S-oxide (40) with acetic anhydride in the presence of an organic acid, to afford the corresponding acetate (41) with 85-90% stereospecificity as shown below (eq. 52).[66]

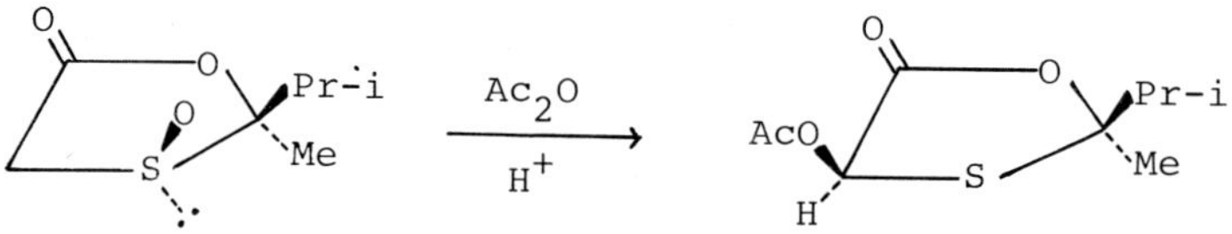

(52)

Another example of the stereospecific Pummerer reaction is the reaction between (S)-3-cephem S-oxide (42) with ethyl chlorocarbonate in the presence of triethylamine to afford the corresponding 2-carbonate ester (43), found by Bremner and Campbell (eq. 53).[67] The (R)-sulfoxide does not react under the same conditions.

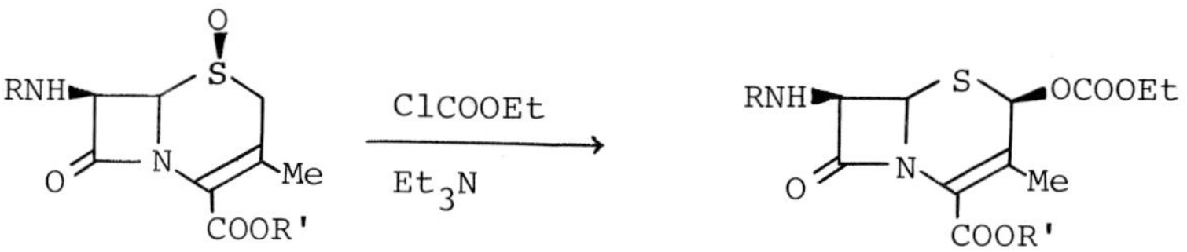

(53)

R = PhOCH₂CO-   R'= Cl₃CCH₂-

What would be the extent of asymmetric induction in the reaction with this rather rigid system ?   No answer has yet been obtained to this question.

McCormick et al. recently reported that the Pummerer reaction of the sulfoxide derivatives of a thiosugar (44) with acetic anhydride also gave the corresponding stereospecific rearrangement products (eqs. 54 and 55).[68]

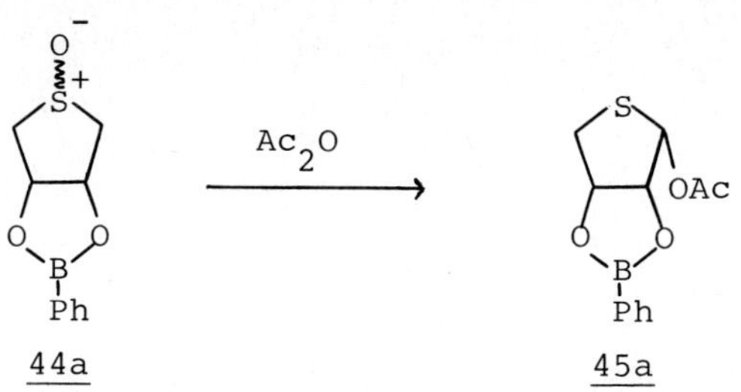

$$\text{44a} \quad \xrightarrow{\text{Ac}_2\text{O}} \quad \text{45a} \tag{54}$$

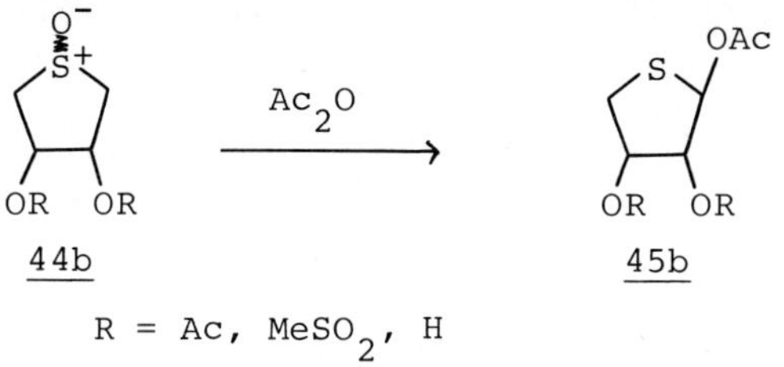

$$\text{44b} \quad \xrightarrow{\text{Ac}_2\text{O}} \quad \text{45b} \tag{55}$$

R = Ac, MeSO$_2$, H

An interesting reaction which is presumed to proceed via an intimate ion pair is the thermal rearrangement of the following α-trimethylsilylsulfoxide (46),[69-72] and a marked conformational dependence was observed in the reaction. Thus, the _cis_-conformer of 2-trimethylsilyl-1,3-dithiane S-oxide was found to undergo rapidly the Pummerer reaction compared to the _trans_-conformer,[71] seemingly suggesting that the reaction proceeds via a four-membered cyclic transition state shown below (eq. 56).

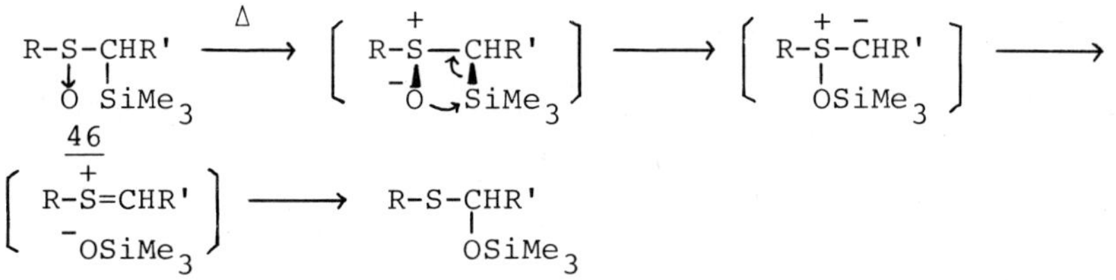

When the following azasulfonium salts were treated with trialkyl-amine around 0°C, the Pummerer reaction, which appeared to be intramolecular, took place as shown in eqs. 57,[73] 58,[74] 59,[75] 60,[76] and 61,[77] though no mechanistic details has been presented.

$$RCH_2-\overset{+}{\underset{\underset{O=C\diagdown_N\diagup C=O}{|}}{S}}-CH_3 \ Cl^- \quad \xrightarrow[72-95\%]{Et_3N, \ 0°} \quad RCH_2-S-CH_2O-\text{(imide)}=O \qquad 9$$

$$+ \ RCH_2-S-CH_2-N\diagup^{CO}_{CO} \qquad 1 \tag{57}$$

$$CH_3-\overset{+}{\underset{\underset{R}{\underset{N}{|}}{\diagdown}COR'}{S}}-CH_3 \ Cl^- \quad \xrightarrow[66-84\%]{Et_3N, \ 0°} \quad R'\overset{O}{\underset{\parallel}{C}}-\overset{R}{\underset{|}{N}}-CH_2-S-CH_3 \tag{58}$$

Eq. (59):
$$R-\overset{+}{S}-CH_3 \ Cl^- \quad \xrightarrow{Et_2NPr-\underline{i}} \quad \begin{array}{c} Me\diagdown \\ Me\diagup \end{array} \text{(ring)} -OCH_2SR \tag{59}$$

Eq. (60):
$$R-\overset{+}{S}-CH_3 \ 2\ Cl^- \quad \xrightarrow{Et_3N} \quad \begin{array}{c} Me \\ Me \end{array}\text{(ring)}-OCH_2SR , \ RSCH_2O-$$

$$+ \quad \begin{array}{c} Me \\ Me \end{array}\text{(ring)}-OCH_2SR, \ CH_2SR \tag{60}$$

$$CH_3-\overset{+}{\underset{\underset{Ar}{\underset{N}{|}}{\diagdown}COMe}{S}}-CH_3 \ Cl^- \quad \xrightarrow{Et_3N} \quad Ar-N=\overset{CH_3}{\underset{|}{C}}-O-CH_2SCH_3 \tag{61}$$

Sulfonium ylides (49)[78] and (50)[79] were reported to undergo
what appears to be an intramolecular Pummerer reaction upon refluxing
with water or ethanol (eqs. 62 and 63). Although both reactions

$$Ph-\overset{O}{\underset{\parallel}{C}}-\overset{-}{C}H-\overset{+}{\underset{\diagdown CH_3}{S}}\diagup^{CH_3} \quad \xrightarrow[H_2O]{refluxing} \quad Ph-\overset{OCH_2-S-CH_3}{\underset{|}{C}}=CH_2 \tag{62}$$

$$\underline{49} \qquad\qquad 66\%$$

$$ArCH=\overset{O}{\underset{\underset{Ph}{\underset{|}{C}}}{\overset{\parallel}{C}}}-\overset{-}{C}H-\overset{+}{\underset{\diagdown CH_3}{S}}\diagup^{CH_3} \quad \xrightarrow[EtOH]{refluxing} \quad ArCH=\overset{OCH_2-S-CH_3}{\underset{\underset{Ph}{\underset{|}{C}}}{\underset{|}{C}}}=CH_2 \tag{63}$$

$$\underline{50} \qquad\qquad 30-62\%$$

can be explained nicely by assuming a five membered cyclic intimate
ion pair at the transition state of the migration, further mechanistic
studies are necessary before we can understand the mechanistic

details. One more example of the Pummerer reaction which is likely
to be intramolecular is the following reaction between β-ketosulfoxide (51) and N-sulfinyltosylamide (eq. 64).[80] One likely

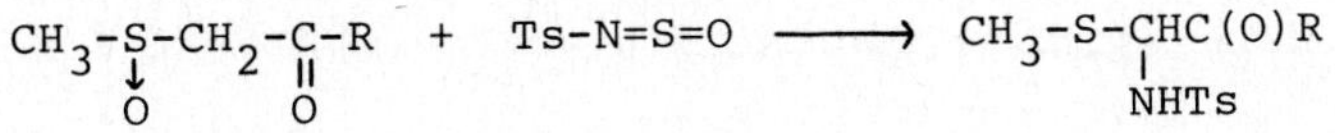

$$CH_3-S-CH_2-C-R \ + \ Ts-N=S=O \longrightarrow CH_3-S-CHC(O)R \tag{64}$$

with O and O below S and C respectively, and NHTs below CHC(O)R.

51

mechanism may be illustrated in the following equation (eq. 65),
which could be supported readily with the aid of both ^{18}O and
deuterium tracer experiments, though these have not been performed
as yet.

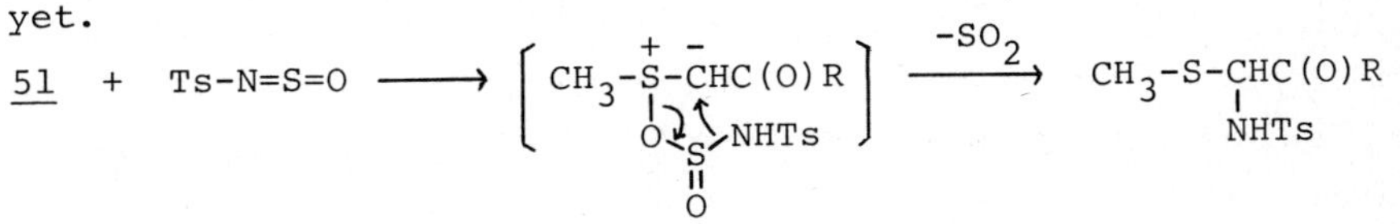

$$51 \ + \ Ts-N=S=O \longrightarrow \left[CH_3-\overset{+}{S}-\overset{-}{CHC}(O)R \right] \overset{-SO_2}{\longrightarrow} CH_3-S-CHC(O)R \tag{65}$$

with O–S(=O)–NHTs bridging structure in brackets, and NHTs below the product.

II. G. Intermolecular Pummerer Reactions

In the early sixties we carried out an ^{18}O tracer experiment on
the Pummerer reaction of dimethyl sulfoxide with uniformly ^{18}O-
labeled acetic anhydride in ether solution and analyzed the ^{18}O-
content of the resulting α-acetoxymethyl methyl sulfide.[7] The
concentration of ^{18}O in the resulting ester was 3/4 of that of the
acetic anhydride used. Thus all the oxygen atoms of the starting
materials, i.e. DMSO and acetic anhydride, were completely
scrambled in the resulting ester (eq. 66). On the basis of the ^{18}O

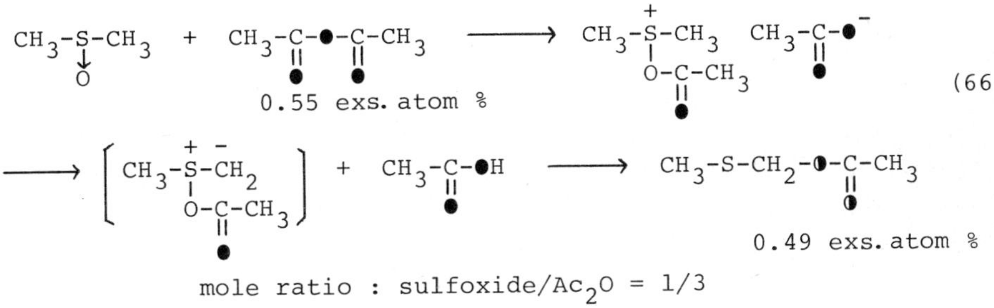

$$CH_3-S-CH_3 \ + \ CH_3-C-\bullet-C-CH_3 \longrightarrow CH_3-\overset{+}{S}-CH_3 \quad CH_3-C-\bullet^-$$

0.55 exs. atom %

$$\longrightarrow \left[CH_3-\overset{+}{S}-\overset{-}{CH_2} \right] \ + \ CH_3-C-\bullet H \longrightarrow CH_3-S-CH_2-\bullet-C-CH_3 \tag{66}$$

0.49 exs. atom %

mole ratio : sulfoxide/Ac₂O = 1/3

tracer experiment, we postulated that intermolecular attack of
acetate at the α-carbon is involved in this Pummerer reaction.
Subsequently, sulfoxides were found to undergo oxygen exchange with
acetic anhydride[29] and various other acids.[81-85] This rather facile
oxygen exchange cast some doubt[10] as to the validity of using
the earlier ^{18}O-tracer experiment to support the intermolecular
acetoxy migration. Therefore, both the rate of oxygen exchange and

that of the Pummerer reaction were measured with methyl p-tolyl
sulfoxide. Then the rate of oxygen exchange at 120°C was estimated
as only 1/6 of that of the Pummerer reaction, indicating that the
earlier postulate of the predominant intermolecular attack of
acetoxy group at the α-carbon terminal is still valid.[27,28] However,
our recent findings of several intramolecular Pummerer reactions
with the aid of ^{18}O-tracer experiments have prompted us to reinvesti-
gate the reaction, using this time ^{18}O-labeled sulfoxide so that
the ^{18}O-distribution in the resulting ester can be determined in
order to evaluate the extent of the intramolecular acetoxy migration.
Our earlier observation and conclusion were found to be quite
acceptable. In the Pummerer reaction of benzyl methyl sulfoxide,
only 5% of [18]O was found to migrate from the original sulfoxide
into the resulting ester, while with methyl phenyl sulfoxide the
Pummerer product retained only 3% of [18]O from the original [18]O
label of the starting material.[44]

Earlier, Johnson and Phillips also carried out a cross-over
experiment shown below (eq. 67) between ^{14}C-labeled and unlabeled
aryl ethyl methoxysulfonium salts.[10] This methoxysulfonium salt,

$$\text{Ar-}\overset{+}{\underset{OCH_3}{S}}\text{-CH}_2\text{CH}_3\ BF_4^- \ + \ \text{Ph-}\overset{+}{\underset{\overset{*}{OCH_3}}{S}}\text{-CH}_2\text{CH}_3\ BF_4^- \quad \xrightarrow[\text{acetone}]{\text{base}}$$

(67)

$$1389\ \text{cpm/mole}$$

$$\text{Ar-S-}\underset{OCH_3}{\overset{*}{CH}}\text{CH}_3 \ + \ \text{Ph-S-}\underset{OCH_3}{\overset{*}{CH}}\text{CH}_3 \qquad\qquad \text{base : 2,6-lutidine}$$

$$423\ \text{cpm/mole} \qquad 939\ \text{cpm/mole} \qquad\qquad * \ : \ {}^{14}\text{C-label}$$

however, would undergo rapid methoxy exchange at the sulfur atom.
Nevertheless, this is another example to indicate the intermolecular
migration in the Pummerer reaction, and this intermolecular mechanism
was supported by the observation that racemic α-methoxybenzyl p-tolyl
sulfide was obtained by treatment of optically active (+)-benzyl
p-tolyl methoxysulfonium salt with NaH in THF.[9] The same optically
active salt undergoes the Pummerer reaction by treatment with
pyridine at room temperature to give the racemic pyridinium salt
in 60-70% yields (eq. 68).[86]

In the late sixties, the Pummerer reaction was generally considered
to involve an intermolecular reaction with either solvent or an
external nucleophile,[19] since the positive charge on sulfur in the
sulfonium salt (52) facilitates the removal of an α-proton to

$$(+) \ \underline{p}\text{-Tol-}\overset{+}{\underset{\underset{OMe}{|}}{S}}\text{-CH}_2\text{Ph BF}_4^{-} \quad \xrightarrow[\text{THF}]{\text{NaH}} \quad (\pm) \ \underline{p}\text{-Tol-}\underset{\underset{OMe}{|}}{S}\text{-CHPh}$$

$$\xrightarrow[\text{CH}_2\text{Cl}_2]{\text{C}_5\text{H}_5\text{N}} \quad (\pm) \ \underline{p}\text{-Tol-S-CHPh BF}_4^{-} \tag{68}$$

yield the ylide (53), which would then react directly with solvent
to afford the α-substituted sulfide (54), or lose an acyloxy ion
to form the sulfur-stabilized carbonium ion (55) that reacts
immediately with solvent to give the final product (54), as shown
in eq. 69.

$$\underset{52}{\overset{+}{\underset{\underset{O}{\overset{\parallel}{|}}}{R\text{-}S\text{-}CH_2R'}}} \quad \xrightarrow{\text{base}} \quad \underset{53}{\overset{+}{\underset{\underset{O}{\overset{\parallel}{|}}}{R\text{-}S\text{-}\overset{-}{C}HR'}}} \quad \longleftrightarrow \quad \overset{+}{\underset{\underset{O}{\overset{\parallel}{|}}}{R\text{-}S\text{=}CHR'}} \tag{69}$$

$$[\ \overset{+}{R\text{-}S\text{=}CHR'} \longleftrightarrow \overset{+}{R\text{-}S\text{-}\overset{-}{C}HR'} \ ] \quad \xrightarrow{X^{-}} \quad \underset{\underset{X}{|}}{R\text{-}S\text{-}CHR'} \ + \ R''COO^{-}$$

$$\underset{55}{} \qquad \qquad \qquad \underset{54}{}$$

$$\text{-R''COO}^{-} \qquad X^{-}$$

Accordingly, several attempts have been performed to obtain clear
evidence to support this carbonium ion formation. One unsuccessful
example is the Pummerer reaction of cyclopropyl phenyl (56), cyclo-
propylcarbinyl phenyl (57) and cyclobutylcarbinyl phenyl (58)
sulfoxides with acetic anhydride. If any dissociated carbonium ion

Ph-S◁ R (R = R'= H, R = R'= Me (cis or trans), R = H, R'= Ph)   Ph-S-CH₂◁   Ph-S-CH₂▱

56   57   58

pair is involved during the Pummerer reactions of these sulfoxides,
all these compounds should give at least partially ring-opened
products which were not, however, found to be formed at all among
the products. Instead, only the normal Pummerer rearrangement
products were obtained nearly quantitatively,[61,87] unlike in the

acetolysis of the corresponding α-chlorocycloalkyl phenyl sulfides which give ring opening products. Probably the acyloxysulfonium ylide (53) undergoes heterolysis to form the sulfur-stabilized carbonium ion (55), which, however, is not a dissociated ion but paired intimately with acylate ion. Therefore, the reaction appears to involve an intramolecular migration. However, when the sulfur-stabilized carbonium ion pair is partially dissociated in somewhat polar media, the carbonium ion (55) does seem to recombine with other nucleophiles than the acylate ion being cleaved from the original ylides (53). One example of such an intermolecular Pummerer reaction may be the following aromatic electrophilic substitution by the Pummerer reaction intermediate (eqs. 70 and 71).[88] Apparently

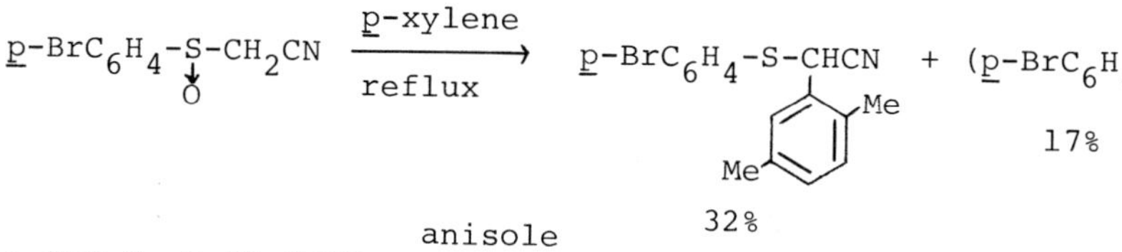

$$\underline{p}\text{-BrC}_6\text{H}_4\text{-}\overset{\downarrow}{\underset{O}{S}}\text{-CH}_2\text{CN} \xrightarrow[\text{reflux}]{\text{p-xylene}} \underline{p}\text{-BrC}_6\text{H}_4\text{-S-CHCN} + (\underline{p}\text{-BrC}_6\text{H}_4\text{S})_2 \quad (70)$$

32%

$$\underline{p}\text{-BrC}_6\text{H}_4\text{-}\overset{\downarrow}{\underset{O}{S}}\text{-CH}_2\text{COOMe} \xrightarrow[\text{reflux}]{\text{anisole}} \underline{p}\text{-BrC}_6\text{H}_4\text{-S-CHCOOMe} + (\underline{p}\text{-BrC}_6\text{H}_4\text{S})_2 \quad (71)$$

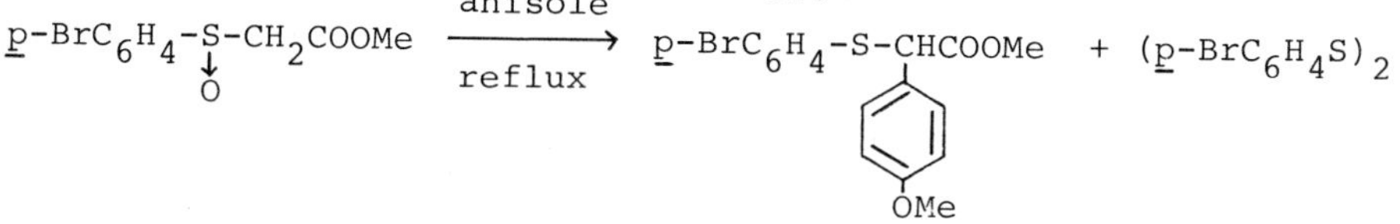

17%

the reaction is autocatalyzed and is presumed to involve the sulfur-stabilized carbonium ion (59)   formed during the reaction (eq. 72).

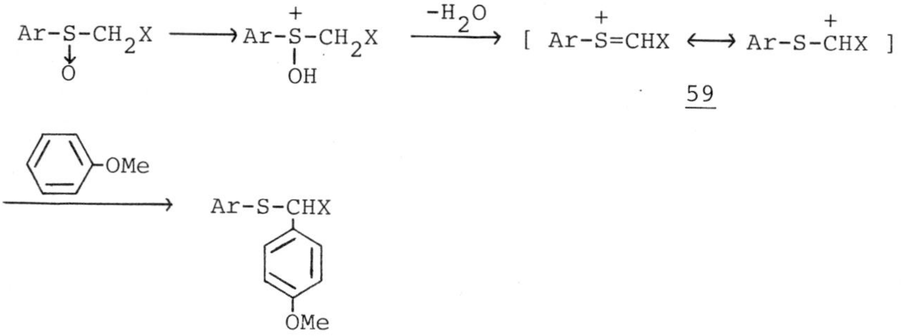

$$\text{Ar-}\overset{\downarrow}{\underset{O}{S}}\text{-CH}_2\text{X} \longrightarrow \text{Ar-}\overset{+}{\underset{OH}{S}}\text{-CH}_2\text{X} \xrightarrow{-\text{H}_2\text{O}} [\ \text{Ar-}\overset{+}{S}\text{=CHX} \longleftrightarrow \text{Ar-S-}\overset{+}{C}\text{HX}\ ]$$

59

$$\xrightarrow{\langle \text{OMe}\rangle} \text{Ar-S-CHX} \quad (72)$$

Phenol or p-substituted thiophenol is a rather powerful trapping agent of carbonium ion and the following reactions are such examples in which both phenol and p-substituted thiophenol successfully trapped a carbonium ion intermediate formed in the reaction of dimethyl sulfoxide with trifluoroacetic anhydride (eqs. 73 and 74).[89] In these reactions the intermediate (59) has a very poor nucleophile, i.e. trifluoroacetoxy anion, as a counter ion, so that they are susceptible to nucleophilic attack by an additional nucleophile.

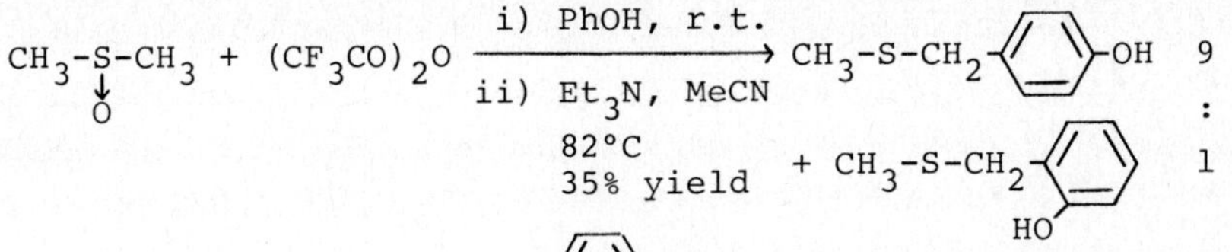

$$CH_3-S-CH_3 + (CF_3CO)_2O \xrightarrow[\text{ii) Et}_3N, \text{ MeCN}]{\text{i) PhOH, r.t.}} CH_3-S-CH_2-\underset{}{\bigcirc}-OH \quad 9 \qquad (73)$$

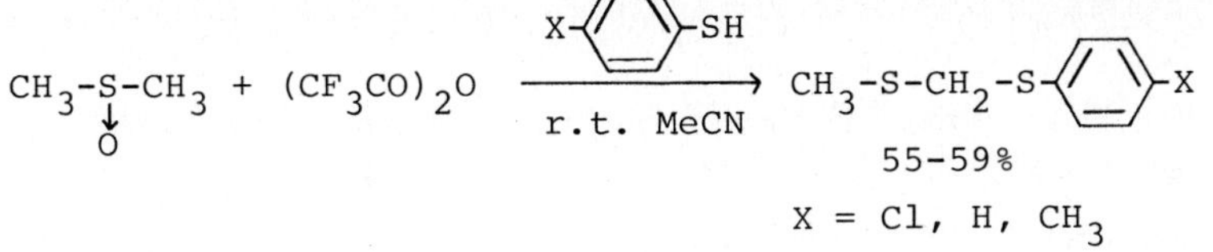

$$\underset{82°C}{\underset{35\% \text{ yield}}{}} \quad + CH_3-S-CH_2-\underset{HO}{\bigcirc} \quad 1$$

$$CH_3-S-CH_3 + (CF_3CO)_2O \xrightarrow[\text{r.t. MeCN}]{X-\bigcirc-SH} CH_3-S-CH_2-S-\bigcirc-X \qquad (74)$$

$$55-59\%$$

$$X = Cl, \ H, \ CH_3$$

In the reaction of the dimethyl ethoxysulfonium salt with triethylamine, o-cresol added to the reaction mixture was found to be
substituted by the intermediate electrophile to give p-methylthio-
methyl substituted o-cresols, though in very low yields (eq. 75).[90]

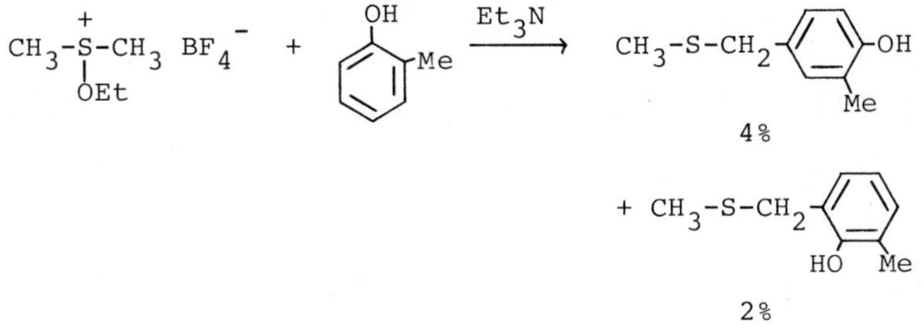

$$CH_3-\overset{+}{S}-CH_3 \ BF_4^- \ + \ \underset{}{\overset{OH}{\bigcirc}}-Me \xrightarrow{Et_3N} CH_3-S-CH_2-\underset{Me}{\bigcirc}-OH$$

$$4\% \qquad (75)$$

$$+ \ CH_3-S-CH_2-\underset{HO \quad Me}{\bigcirc}$$

$$2\%$$

When trans-1,4-dithiane disulfoxide (60) was treated with acetic
anhydride, the ring contracted product (61) was found to be formed
among several products, indicating that the carbonium ion intermediate
(62), stabilized by both α and β sulfur atoms, is involved during
the reaction (eq. 76).[91]

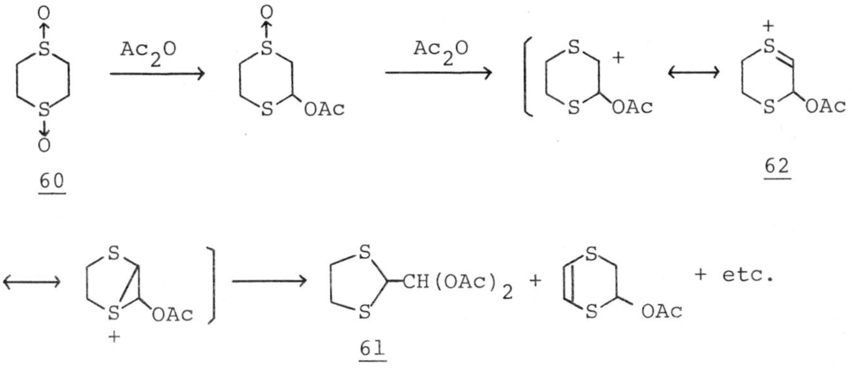

$$(76)$$

The following β-ketosulfoxide undergoes the Pummerer reaction in dilute hydrochloric acid, presumably via formation of the corresponding sulfur-stabilized carbonium ion intermediate (eqs. 77 and 78).[92]

$$CH_3-\overset{\downarrow}{\underset{O}{S}}-CH_2-\overset{O}{\underset{\|}{C}}-Ph \xrightarrow{H_3O^+} CH_3-S-\underset{OH}{\overset{|}{C}}H-\overset{O}{\underset{\|}{C}}-Ph \xrightarrow{H_2O} CH_3SH + Ph-\overset{O}{\underset{\|}{C}}-CHO \quad (77)$$

The rate of the reaction can be nicely correlated with the Bunnett-Olesen treatment and a w-value of +6.7 and φ-value of +1.1 were obtained. These values suggest that water molecules act as proton-abstracting species in the rate-determining step. The size of the activation parameters (E_a = 20.3 kcal/mole, $\Delta S^{\ddagger}$ = -17.3 e.u.), and the small Hammett ρ value (ρ = +0.22) obtained with substituted phenacyl methyl sulfoxides, seem to indicate that the reaction proceeds through the initial fast protonation of sulfinyl oxygen, followed by the rate-determining proton removal with water, and the effect of the substituent is cancelled by protonation at the sulfinyl oxygen and deprotonation of the α-methylene proton. The hydroxyl group of the product was found to be derived exclusively from the solvent, water, in view of the ^{18}O-tracer experiment using ^{18}O-labeled water as solvent. Unfortunately in this case a kinetic isotope effect could not be measured because rapid H-D exchange of the methylene protons occurred by keto-enol isomerization prior to the Pummerer reaction (eq. 78).

$$CH_3-\overset{\downarrow}{\underset{O}{S}}-CH_2-\overset{O}{\underset{\|}{C}}-Ph \rightleftharpoons CH_3-\overset{+}{\underset{OH}{S}}-CH-\overset{O}{\underset{\|}{C}}-Ph \xrightarrow{slow} [\ CH_3-\overset{+}{S}=CH-\overset{O}{\underset{\|}{C}}-Ph$$

$$\longleftrightarrow CH_3-\overset{+}{S}-CH-\overset{O}{\underset{\|}{C}}-Ph\ ] \xrightarrow{H_2O^{18}} CH_3-S-\underset{^{18}OH}{\overset{|}{C}}H-\overset{O}{\underset{\|}{C}}-Ph \quad (78)$$

Recently we found on the basis of a tracer study with ^{18}O-labeled sulfoxides that the Pummerer reaction of such conformationally fixed six-membered sulfur heterocycles as 4-p-chlorophenylthiane S-oxide (cis and trans) proceeds intermolecularly to yield the product ester containing very little ^{18}O of the original sulfoxide, while the recovered sulfoxide was found to have retained the original ^{18}O label completely. However, reactions of both the cis- and the trans- sulfoxides proceeded stereoselectively to give the axial α-acetoxysulfide (66) predominantly on heating with acetic anhydride

alone, while the equatorial α-acetoxysulfide (64) was found to be
formed stereoselectively on heating with acetic anhydride in the
presence of excess of DCC or 2,6-lutidine as an acid-scavenger,
indicating that an acid-catalyzed isomerization of the equatorial
isomer (64) to the thermodynamically stable axial isomer (66) is
involved during the reaction in the absence of the acid-scavenger
(eqs. 79 and 80).[93]

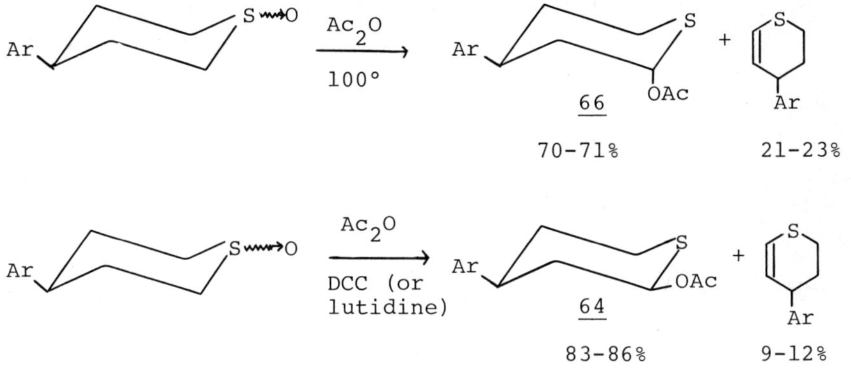

$$\text{Ar} : \text{p-ClC}_6\text{H}_4$$

The intermolecular nature of the acetoxy migration, despite the
lack of oxygen exchange in the sulfoxide, seems to suggest that
the reaction proceeds through the formation of a sulfur-stabilized
sp^2 carbonium ion intermediate (63), which is attacked by external
acetoxy group preferentially from the upper direction (path a) to
give the equatorial α-acetoxysulfide (64), while attack from the
bottom side (path b) gives the boat form α-acetoxysulfide (65).
The latter would be less favorable in conformational stability,
and eventually undergo conformational change to give the axial
isomer (66) (eq. 80). In this reaction, however, the rate-determining
step is proton removal in view of the relatively large isotope
effects (k_H/k_D = 2.8 for the cis isomer, and 3.4 for the trans
isomer).

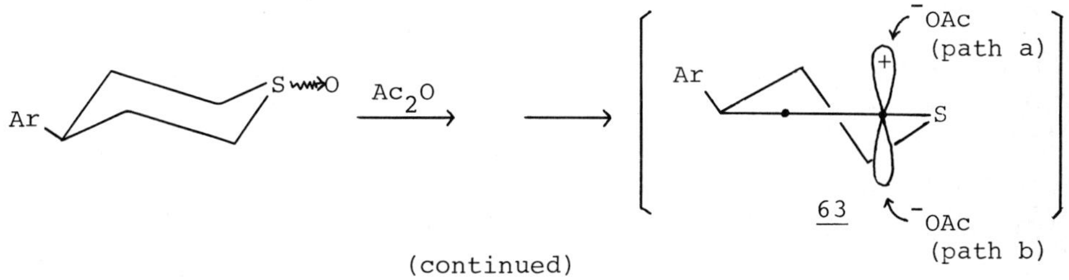

(continued)

(80)

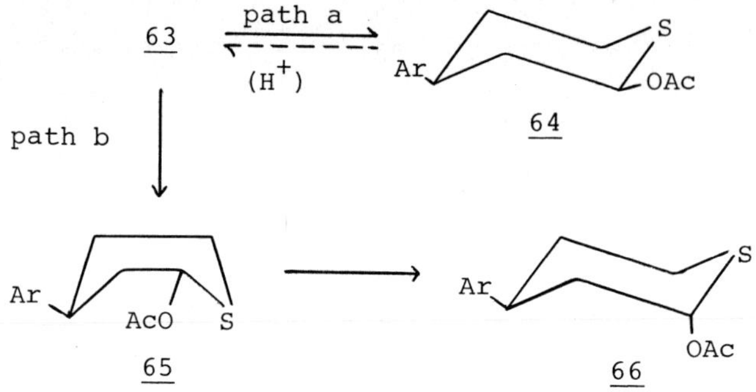

$$63 \underset{(H^+)}{\overset{\text{path a}}{\rightleftharpoons}}$$

path b

64

Ar / AcO S

65

→ Ar S OAc

66

The following allylic sulfoxide (67) was reported to afford the two γ-acetoxy sulfides in good yields in the ratio of 3 : 1, apparently through an intermolecular route via an intermediate (68) as shown below.[94]

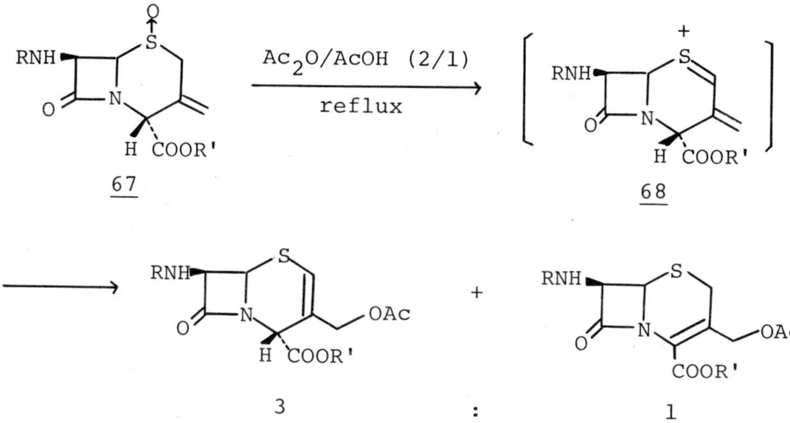

$$\xrightarrow[\text{reflux}]{Ac_2O/AcOH \ (2/1)}$$

67   68   (81)

→   RNH ... OAc   +   RNH ... OAc

3   :   1

The base-catalyzed rearrangement reaction of the following thioxanthene sulfimide (69) may be classified as an intermolecular Pummerer type of reaction (path a). In this case an intramolecular 1,4-sigmatropic rearrangement mechanism (path b) is less likely, although it cannot be ruled out completely (eq. 82).[95]

In connection of the [18]O tracer experiments which we have conducted for mechanistic investigations, it would be appropriate to mention briefly the usual method of [18]O analysis that was used. Basically, it is an application of the old method of Rittenberg and Ponticorvo.[96] Thus the sample to be analyzed was incinerated together with a mixture of $HgCl_2$ and $Hg(CN)_2$ at 450–500°C for a few hours, the CO_2 gas formed was removed and purified for direct mass spectrometric

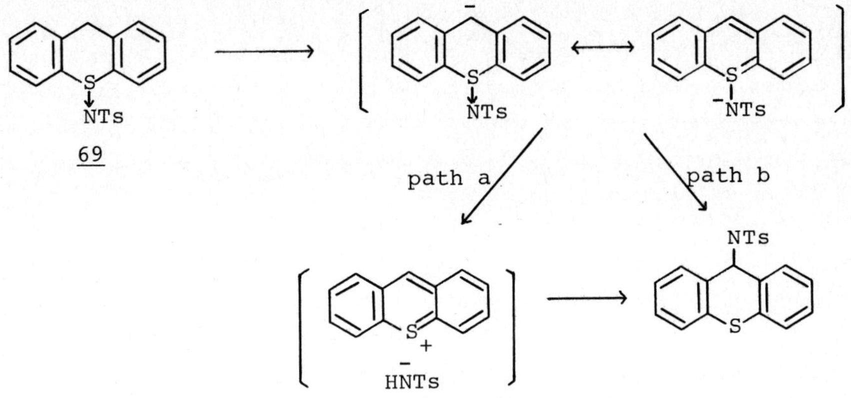

$$(82)$$

analysis, and then from the mass ratio of 46/44, the content of [18]O was estimated. After more than a decade of trials and errors, we have found that the best analytical results can be obtained when a 1 : 1 mixture of $HgCl_2$, purified by repeated sublimation, and $Hg(CN)_2$, recrystallized from absolute ethanol, was used in about 30 fold excess over the sample to be analyzed. For routine analyses, pyrex glass tubing suffices but for precise [18]O measurements quartz glass tubing is preferable, the error being confined within ± 0.6%.[97]

III.   OTHER PUMMERER TYPE REACTIONS

III. A.   Reaction between Sulfides and Halogenating Agents

Sulfides are known to form addition complexes upon treatment with elemental halogen, and the sulfides bearing an α-methylene hydrogen usually react further to form α-halosulfides. This reaction can be classified as the Pummerer reaction of halosulfonium salts (70) (eq. 83). Not only chlorine[98-101] and bromine[100,102], but also sulfuryl chloride,[99,103-107] 3-iodopyridine - chlorine complex,[108] N-chlorosuccinimide[109-116] and N-bromosuccinimide[113,117] can serve as effective halogenating agents and afford α-halosulfides via the Pummerer type reaction of halosulfonium salts. Sometimes halogenation continues further to give di- or tri- halogenosulfide depending upon the reaction condition.[99,101,104-107]

$$R-S-CH_3 + X_2 \longrightarrow \overset{+}{R-\underset{X}{S}}-CH_3 \ X^- \longrightarrow \overset{+}{R-\underset{X}{S}}-CH_2^- + HX \longrightarrow R-S-CH_2X \quad (83)$$

$$\underline{70} \qquad\qquad \underline{71}$$

There have been a few studies which give rather conclusive evidence that the rate-determining step is the proton removal. For example, Tuleen and Marcum showed that in the chlorination of α-d_1-benzyl phenyl sulfide with NCS, the kinetic isotope effect was estimated to be about 5.3-5.9 from the observation of internal competitive chlorination (eq. 84).[112]

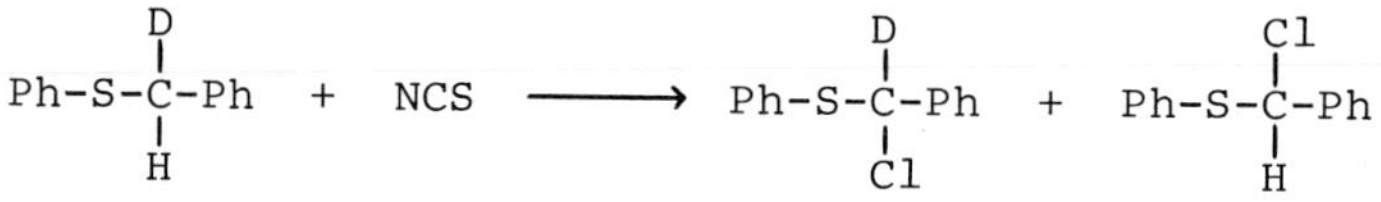

$$
\underset{\overset{|}{H}}{\overset{\overset{D}{|}}{Ph-S-C-Ph}} + NCS \longrightarrow \underset{\overset{|}{Cl}}{\overset{\overset{D}{|}}{Ph-S-C-Ph}} + \underset{\overset{|}{H}}{\overset{\overset{Cl}{|}}{Ph-S-C-Ph}}
\tag{84}
$$

Wilson, Jr. and Albert estimated the isotope effect (k_H/k_D) to be 5.1 with chlorine, and 3.6 with bromine, in the intramolecular competitive reaction of α,α-d_2-thiolane with chlorine or bromine in methanol to form α-halothiolane (eq. 85).[100] The complex between

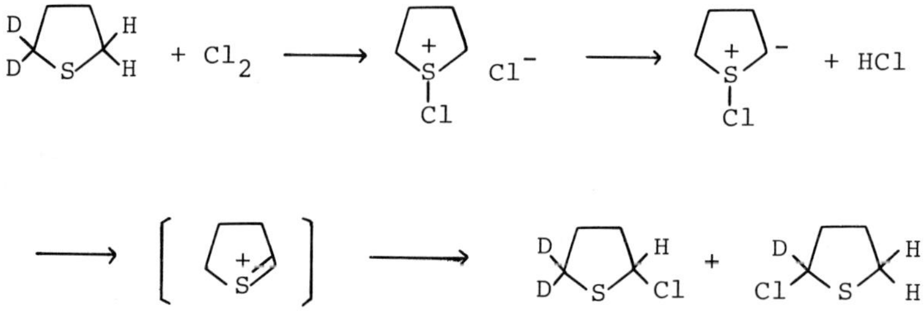

$$\tag{85}$$

thiolane and bromine is not a sulfonium salt but a charge-transfer complex (72) in the crystalline state, as shown below,[118] but would be ionized in such a polar solvent as methanol. Therefore, the

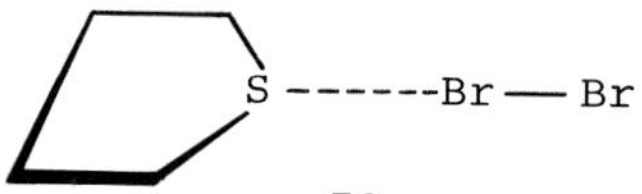

72

reaction undoubtedly occurs between the sulfonium ion and halide ion and proceeds via formation of the halosulfonium ylide. Involvement of the halosulfonium ylide was confirmed by a recent spectroscopic identification of the ylide in the reaction of 1,3-dithiane with sulfuryl chloride (eq. 86).[119]

In a special case, a sulfur-stabilized carbonium ion type compound was isolated in the reaction of the sulfide (73) with sulfuryl chloride; the thiapyrylium salt (74) produced is stabilized by

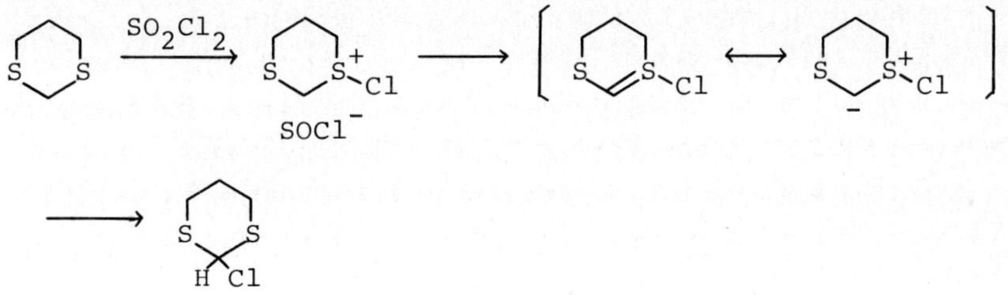

(86)

aromatic conjugation (eq. 87).[120]

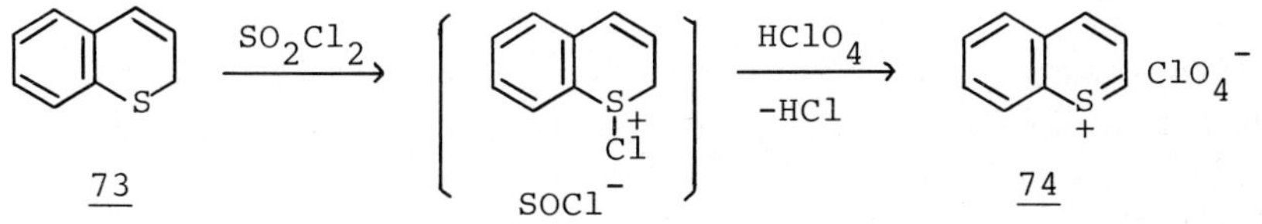

(87)

Thus, the initial formation of the sulfonium ion (73) and the subsequent proton removal to form the halosulfonium ylide (71) are beyond doubt analogous to the usual Pummerer reaction of sulfoxides. How does the rearrangement of halogen from sulfur to α-carbon proceed, intramolecularly or intermolecularly ? Unfortunately there is no experimental work to substantiate the argument for either case, nor would it be easy to determine the nature of the halogen migration experimentally.

One interesting problem is the regioselectivity of the Pummerer type α-halogenation of sulfides. A few experimental data have revealed that the migration of halogen takes place at the α-carbon which has more alkyl substituents, contrary to observation in the usual Pummerer reaction of sulfoxides with acetic anhydride (vide supra Section II. C.). An example is shown below (eq. 88).[115] In

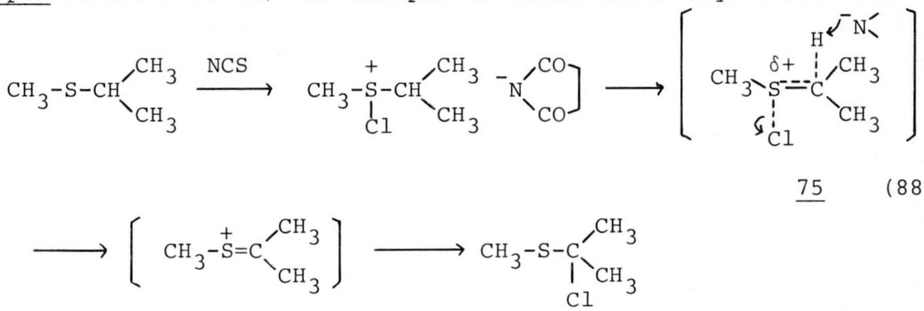

(88)

this case, heterolytic cleavage of the S-Cl bond would be easy, while the nucleophile to remove the α-proton is a weak acylimide base.[121] Therefore, the situation is somewhat like the Saytzeff type concerted elimination of alkyl halide with bases. If we assume a Saytzeff type E2 elimination in this case, the transition state (75) should be stabilized with more α-alkyl substituents, since both the π-conjugation between the C-S bond and a partial positive charge are developed at the α-carbon in the transition state.

In the reaction of sulfoxides with hydrochloric acid in the presence of molecular sieve to afford α-chlorosulfides, chlorination takes place preferentially at the α-carbon with more alkyl substituents.[122] The situation is quite analogous, since the initial step of the reaction between the sulfoxide and hydrochloric acid is presumably the formation of the corresponding chlorosulfonium ion[85] which can undergo the Saytzeff type E2 elimination.

There are many other examples of α-halogenation of sulfoxides, through the Pummerer reaction with various halogenating agents. Various carboxylic acid chlorides,[103,110,123,124] a few sulfenyl,[125,126] sulfinyl,[127] and sulfonyl[128-130] chlorides, thionyl chloride,[103,110,131,132] boron trichloride,[133] chlorosilanes,[133-135] and phosphorus and phosphoryl chlorides[123,131,136,137] have been used successfully. However, not all these acid chlorides are good α-halogenating agents. Sometimes acid chlorides act as reducing agents to deoxygenate sulfoxides to the corresponding sulfides. Among these, acetyl chloride is a good reducing agent for various sulfoxides even at room temperature.[138] The following mechanism was suggested for the reaction (eq. 89).

$$
\begin{array}{ccc}
\underset{\underset{O}{\downarrow}}{R-S-R'} + \underset{\underset{O}{\parallel}}{Me-C-Cl} & \longrightarrow & \overset{+}{\underset{\underset{OCOMe}{|}}{R-S-R'}}\ Cl^- \longrightarrow \overset{+}{\underset{\underset{Cl}{|}}{R-S-R'}}\ MeCOO^-
\end{array} \tag{89}
$$

$$
\xrightarrow{\text{MeCOCl}}\ R-S-R' + Cl_2 + (MeCO)_2O
$$

Recently, sulfoxides were found to be rapidly reduced to the corresponding sulfides by iodide ion,[139] H_2S[140] and Me_2S[141] after activating the sulfoxides with trifluoroacetic anhydride. Yields are nearly quantitative and the reaction conditions are quite mild (0°C with iodide ion and Me_2S, and -60°C with H_2S). With dimethyl sulfide, not only the reduction but also the intermolecular Pummerer reaction of dimethyl sulfide took place, though the mechanism has not been fully elucidated (eq. 90).[141]

The reaction between sulfoxides and sulfinyl chlorides does not seem to lead to the Pummerer reaction but results in the reduction of sulfoxides to the corresponding sulfides.[129,142] Dialkyl, alkyl aryl, diaryl and benzylic sulfoxides are smoothly reduced by both methanesulfinyl and arenesulfinyl chlorides (eq. 91).[142]

This is a clear-cut oxygen transfer reaction. However, it is not clear whether the reduction proceeds by one step oxygen transfer, or via the formation of a covalent type intermediate (76), or through the consecutive formation of the two ionic sulfonium intermediates (77) and (78) (eq. 92).

Various sulfoxides were reported to be reduced to the corresponding sulfides by treatment with thionyl chloride.[143] Recently, however, treatment of thiolane S-oxide with thionyl chloride was reported to give not only the reduced sulfide but also a small portion of α-chlorothiolane, clearly the Pummerer type rearrangement product (eq. 93)[144]

While methyl phenyl, benzyl phenyl and diphenyl N-tosylsulfimides can be reduced readily to the corresponding sulfides by treatment with acetyl chloride,[138] the reaction of dimethyl N-acetylsulfimide with acetyl chloride or thionyl chloride is reported to afford the α-chlorosulfide, the Pummerer type reaction product.[145]

III. B. Reactions of Thiolsulfinates with Acetic Anhydride

Thiolsulfinates (79) belong to the sulfoxide family in a broad sense. However, unlike other trivalent organic sulfur species such

$$RCH_2\text{-}\overset{\downarrow}{\underset{O}{S}}\text{-}S\text{-}R' \qquad \underline{79}$$

as sulfimides or sulfonium ylides, very little has been known about the reaction between thiolsulfinates with electrophiles. There is a report, however, on the reaction between a cyclic thiolsulfinate, i.e. β-lipoic acid (80)(monooxide of lipoic acid), and acetic anhydride to result in the formation of the Pummerer reaction product (81), though in a very low yield (6%) (eq. 94).[146]

$$\text{(94)}$$

$$\underline{80} \qquad\qquad\qquad\qquad \underline{81}$$

Two other examples of the Pummerer type reactions are with open-chain thiolsulfinates as shown in eq. 95,[147] and eqs. 96 and 97.[148]

$$RCH_2CH_2\text{-}\overset{\downarrow}{\underset{O}{S}}\text{-}S\text{-}CH_2CH_2OCH_3 \xrightarrow[90°C]{MeOH} RCH_2\underset{OMe}{\overset{|}{CH}}\text{-}S\text{-}S\text{-}CH_2CH_2OCH_3 \qquad \text{(95)}$$

R = MeO, H, MeCOO

90-97%

$$CH_3\text{-}\overset{\downarrow}{\underset{O}{S}}\text{-}S\text{-}CH_3 \xrightarrow[\text{reflux, 52 hr}]{C_6H_6\text{-}H_2O(1\text{ eq.})} CH_3\text{-}\overset{\downarrow}{\underset{O}{S}}\text{-}CH_2\text{-}S\text{-}S\text{-}CH_3 \qquad \text{(96)}$$

84%

$$CH_3CH_2\text{-}\overset{\downarrow}{\underset{O}{S}}\text{-}S\text{-}CH_3 \xrightarrow[\text{reflux, 23 hr}]{C_6H_6\text{-}H_2O(1\text{ eq.})} CH_3CH_2S(O)CH_2SSCH_3 \ (66.5\%) \quad \text{(97)}$$

$$+ \ CH_3CH_2S(O)CH_2SSCH_2CH_3 \ (22.2\%)$$

$$+ \ CH_3CH_2S(O)CH(CH_3)SSCH_3 \ (3.1\%)$$

Although the mechanism of the former reaction has not been explored, the latter reaction has been studied extensively by Block and co-workers.[148] They suggested that the reaction proceeded through an incipient formation of sulfinyl sulfonium intermediate (82), formed by the reaction of the starting thiolsulfinate with either the sulfenic acid or the protonated thiolsulfinate, followed by intramolecular hydrogen abstraction to afford the corresponding sulfenic acid and the sulfur-stabilized cationic intermediate (83), both of which recombined to give the α-sulfinyldisulfide (84) (eq. 98).[148] This reaction was suggested to be most favorable

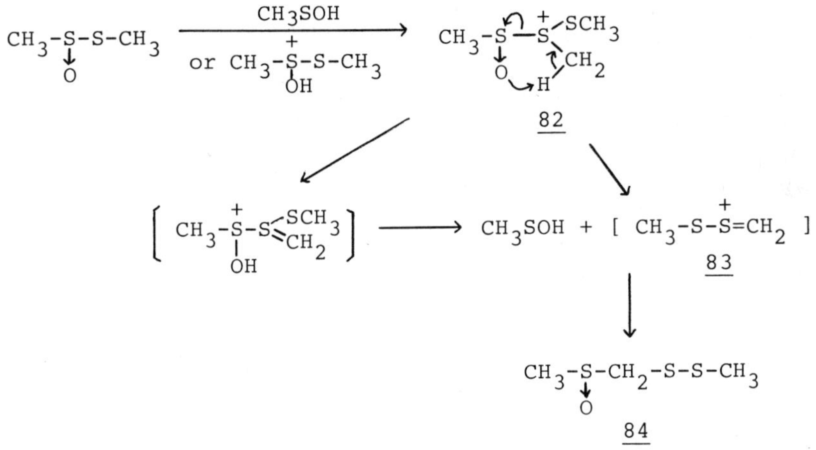

in the presence of an equimolar amount of water, while under anhydrous conditions the α-sulfinyldisulfide became a minor product and the α-sulfonyldisulfide was formed preferentially by the reaction of (83) with the sulfinic acid formed during the reaction.

We found recently that dibenzyl thiolsulfinate (85) reacted with acetic anhydride in the presence of an equimolar amount of acetic acid to afford the rearrangement product (86), consisting of a nearly equimolar mixture of threo and erythro isomers, via a route somewhat different from the Pummerer reaction (eq. 99).[149] In order

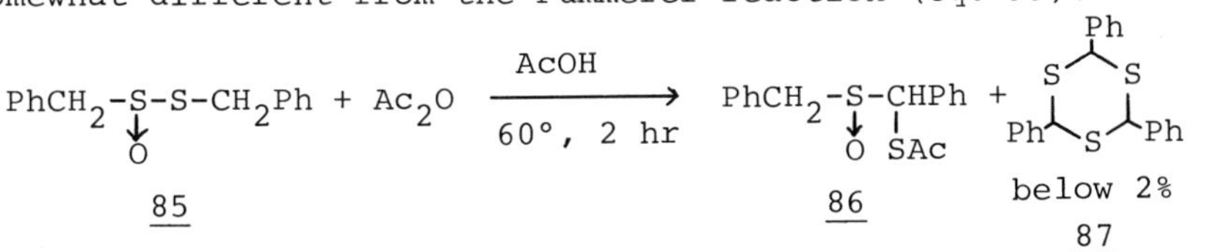

to examine the mechanism of the reaction, both D-labeled and ^{13}C-labeled dibenzyl thiolsulfinates (sulfenyl methylene-D and ^{13}C) were synthesized and subjected to the reaction. Both the product (86') and the recovered thiolsulfinate (85") were found to have lost both D and ^{13}C considerably as shown below (eq. 100). Meanwhile, the ^{18}O-label in the original thiolsulfinate (85') was retained completely both in the product (86') and the recovered thiolsulfinate (85"). Incidentally, there was no loss of deuterium at the sulfinyl

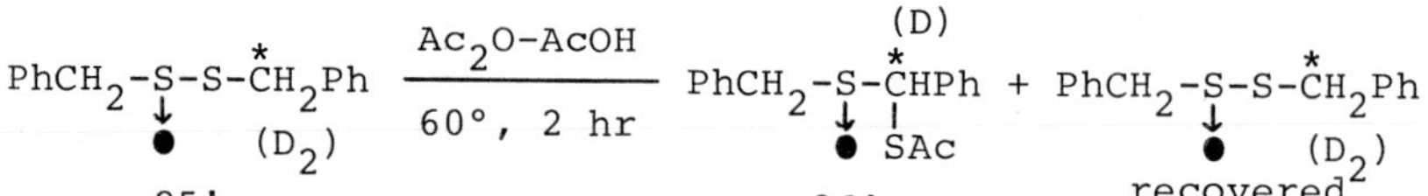

$$\underset{\underset{85'}{\overset{\bullet}{(D_2)}}}{PhCH_2\text{-}S\text{-}S\text{-}\overset{*}{C}H_2Ph} \xrightarrow[60°,\ 2\ hr]{Ac_2O\text{-}AcOH} \underset{\underset{86'}{\overset{\bullet}{SAc}}}{PhCH_2\text{-}S\text{-}\overset{\overset{(D)}{*}}{C}HPh} + \underset{\underset{\substack{recovered \\ 85''}}{\overset{\bullet}{(D_2)}}}{PhCH_2\text{-}S\text{-}S\text{-}\overset{*}{C}H_2Ph}$$

(100)

	relative content (%)		
D	100	75	88
^{13}C (*)	100	76-78	60-62
^{18}O (•)	100	100	100

methylene group in the recovered thiolsulfinate when the thiolsulfinate - sulfinyl methylene-d$_2$ (85"') was used.

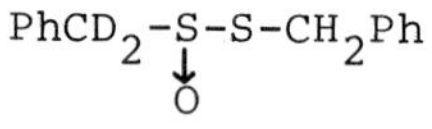

$$\underset{85'''}{PhCD_2\text{-}S\text{-}S\text{-}CH_2Ph}$$
(with O on the first S)

These experimental observations together with the isolation of a small amount of the trithiane (87), a cyclic trimer of thiobenzaldehyde, could be rationalized by assuming the initial dissociation of the thiolsulfinate, the subsequent recombination and the final acetylation to give the rearrangement product (86), with concomitant recombination of the sulfenic acid (88) to give the thiolsulfinate (85) (eq. 101).[150]

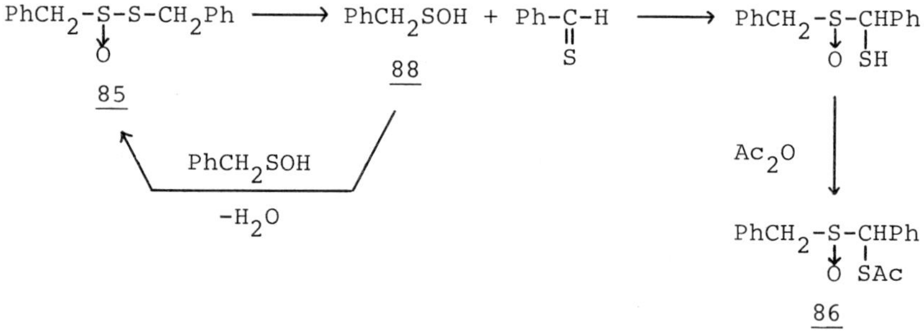

$$\underset{85}{PhCH_2\text{-}\overset{O}{S}\text{-}S\text{-}CH_2Ph} \longrightarrow \underset{88}{PhCH_2SOH} + \underset{S}{Ph\text{-}\overset{\parallel}{C}\text{-}H} \longrightarrow \underset{O\ SH}{PhCH_2\text{-}S\text{-}CHPh}$$

PhCH$_2$SOH

-H$_2$O

Ac$_2$O

$$\underset{86}{PhCH_2\text{-}S\text{-}CHPh}$$ (O SAc)

(101)

III. C. Reaction between Sulfides and Peroxides

There is an example of the Pummerer reaction between organic
sulfides and benzoyl peroxide resulting in the formation of both
the rearranged benzoate ester and the corresponding sulfoxide
(eq. 102).[151-155] Undoubtedly the initial step involves nucleophilic

$$R-S-CH_3 + Ph-\underset{\underset{O}{\|}}{C}-O-O-\underset{\underset{O}{\|}}{C}-Ph \longrightarrow R-S-CH_2OCOPh + PhCOOH \qquad (102)$$

$$+ R-\underset{\underset{O}{\downarrow}}{S}-CH_3 + (PhCO)_2O$$

attack of the sulfide on the O-O bond of the benzoyl peroxide to
form the benzoyloxysulfonium salt, the same intermediate as in the
normal Pummerer reaction.

Another example is the acid-catalyzed oxidation of substituted
phenylmercaptoacetic acid with hydrogen peroxide as shown below,
which is also presumed to proceed via the formation of the sulfoxide,
eventually yielding substituted thiophenol (eq. 103).[156]

$$Ar-S-CH_2COOH \xrightarrow{H_2O_2, \ H^+} \left[Ar-\underset{\underset{O}{\downarrow}}{S}-CH_2COOH \right] \xrightarrow{H^+}$$

$$(103)$$

$$\left[Ar-\underset{\underset{OH}{|}}{S}-CHCOOH \right] \xrightarrow{H^+} ArSH + OHCCOOH$$

III. D. Reaction between Alkyl sulfoxides and Grignard Reagents

When methyl p-tolyl sulfoxide was treated with phenyl Grignard
reagent, benzyl p-tolyl sulfide was reported to be obtained in
good yield (eq. 104).[157]

$$p-Tol-\underset{\underset{O}{\downarrow}}{S}-CH_3 + PhMgX \longrightarrow p-Tol-S-CH_2Ph \qquad (104)$$

Treatment of dimethyl sulfoxide with Grignard reagents, RMgX,
also gave the corresponding sulfides, RCH_2-S-CH_3.[158] The following
mechanism, which involves the oxysulfonium ylide intermediate (89),
has been suggested (eq. 105),[158] although no mechanistic investigation

$$R-\underset{\underset{O}{\downarrow}}{S}-CH_3 + R'MgX \longrightarrow \left[R-\overset{+}{\underset{\underset{OMgX}{|}}{S}}-\overset{-}{CH_2} \right] \xrightarrow{R'MgX} R-S-CH_2R' \qquad (105)$$

89

has been carried out on this reaction. Yet the reaction has been utilized for syntheses of aromatic aldehydes (eq. 106).[159]

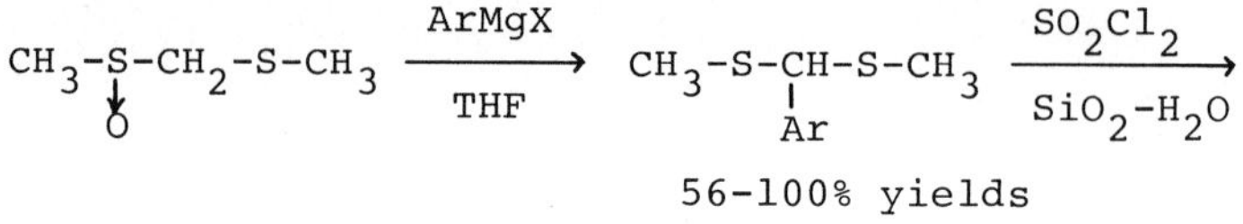

$$CH_3-S-CH_2-S-CH_3 \xrightarrow[\text{THF}]{\text{ArMgX}} CH_3-S-CH-S-CH_3 \xrightarrow[\text{SiO}_2-H_2O]{\text{SO}_2Cl_2} ArCHO \qquad (106)$$

with the first reagent bearing O on the first S, and the product bearing Ar on the central CH.

56-100% yields

III. E.   The Pummerer Type Addition Reactions

Dithioacetic acid is a fairly strong acid (pKa = 2.55) and yet has a strong nucleophilic thiol group. Therefore, it can add to nucleophilic olefins following the Markownikoff rule, and also undergo the Michael type addition toward electrophilic olefins readily without any additive. It is a strong reducing agent and reduces sulfoxides, sulfimides and sulfonium ylides back to the original sulfides even under cold conditions.[160] When phenyl vinyl sulfoxide was added to dithioacetic acid (2 mole) at room temperature, an exothermic reaction took place and 1,2-bis(dithioacetoxy)ethyl phenyl sulfide (90) was obtained quantitatively (eq. 107).[160]

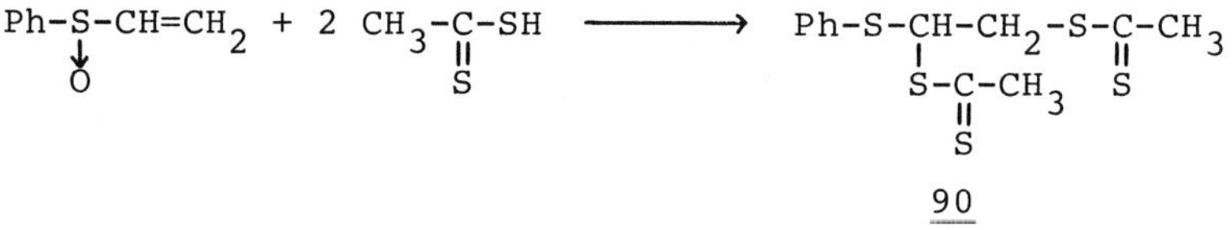

$$Ph-S-CH=CH_2 + 2\ CH_3-C-SH \longrightarrow Ph-S-CH-CH_2-S-C-CH_3 \qquad (107)$$

90

This different pattern of product formation from those of other addition reactions of dithioacetic acids was taken to suggest that the reaction of phenyl vinyl sulfoxide with dithioacetic acid proceeds via initial protonation on the sulfoxide oxygen to form a sulfonium salt (91), followed by addition of dithioacetate to give the Pummerer type intermediate (92) which is eventually attacked by another molecule of dithioacetic acid to yield the final product as shown below (eq. 108).[160]

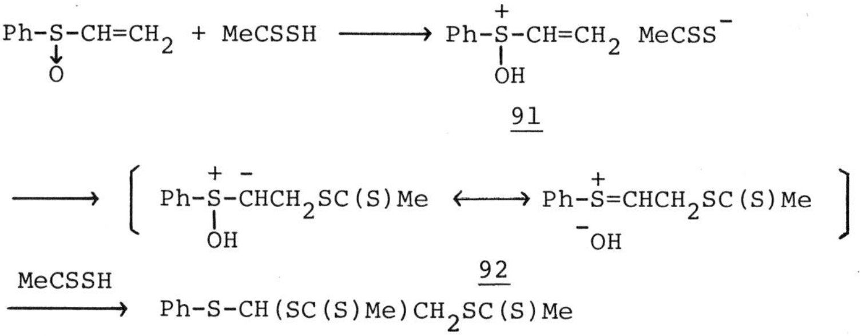

$$Ph-S-CH=CH_2 + MeCSSH \longrightarrow Ph-\overset{+}{S}-CH=CH_2 \ MeCSS^-$$

91

$$(108)$$

$$\longrightarrow \left[Ph-\overset{+}{S}-\overset{-}{C}HCH_2SC(S)Me \longleftrightarrow Ph-\overset{+}{S}=CHCH_2SC(S)Me \right]$$

92

$$\xrightarrow{MeCSSH} Ph-S-CH(SC(S)Me)CH_2SC(S)Me$$

Similar Pummerer type addition reactions are known. The vinyl sulfoxide (93) was reported to react with hydrochloric acid in alcohol according to (eq. 109),[161] and the other vinyl sulfoxide (94) was also shown to react with acetyl chloride to afford the addition product (95) (eq. 110).[162] Another example with vinyl sulfoxide (96) to give the adduct (97) is shown below (eq. 111).[163]

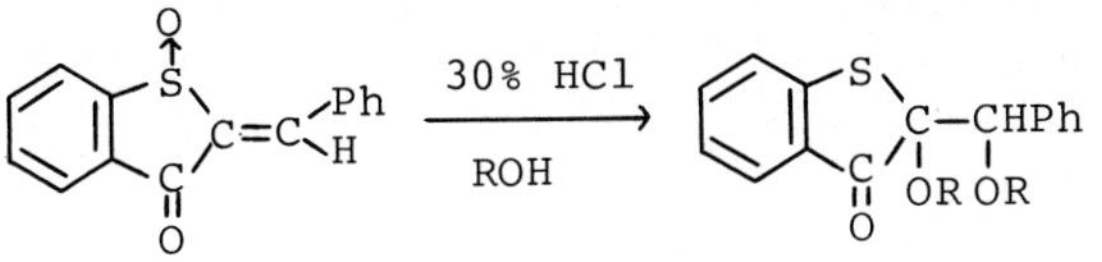

$$(109)$$

93

R = Et   one diastereomer
R = Me   two diastereomers (57 : 25)

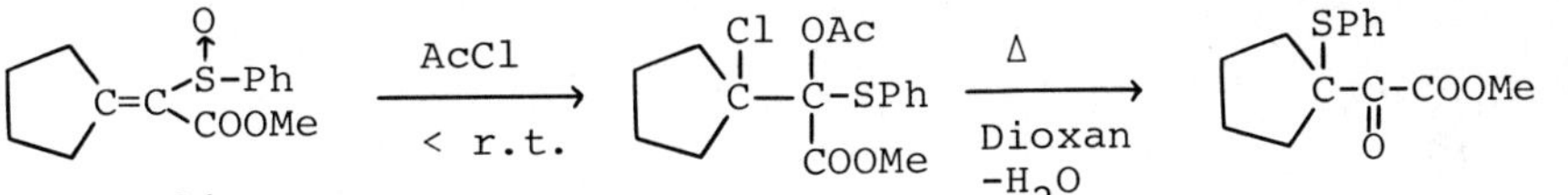

$$(110)$$

94   95

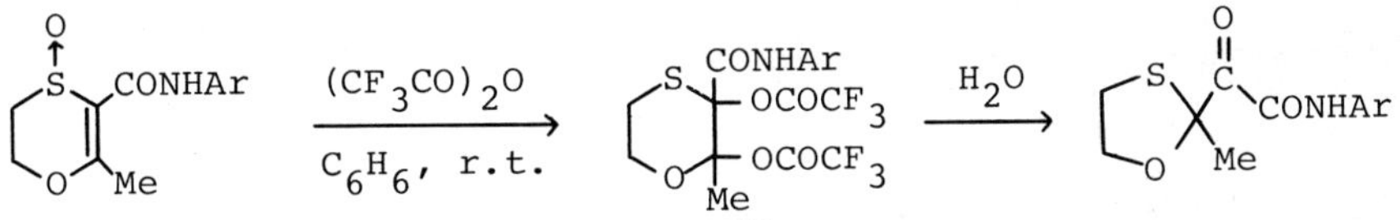

$$(111)$$

96   97

The addition of methoxide to the phenyl vinyl methoxysulfonium salt was found to result in the formation of the Pummerer type product, into which 8.3% deuterium atoms was found to be incorporated on treatment of the methoxysulfonium salt in methanol-d_1. This means that protonation of the ylide intermediate (98) to afford the β-substituted sulfonium salt (99) is involved as a side reaction in this system. (eq. 112).[9]

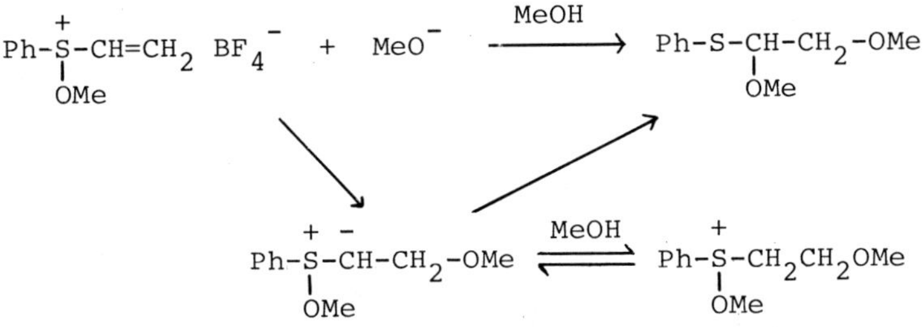

$$(112)$$

98   99

III. F. Abnormal Pummerer Reaction

Morin et al. found that the treatment of penicillin sulfoxides
(100) with acetic anhydride under refluxing gave the ring-expanded
cepham derivatives (103) (eq. 113).[164] This transformation is called
an abnormal Pummerer reaction, signifying that the reaction mechanism
is different from that of the usual Pummerer reaction, involving a
sulfenic acid (101) or sulfenic acid derivative (102) as an intermediate.

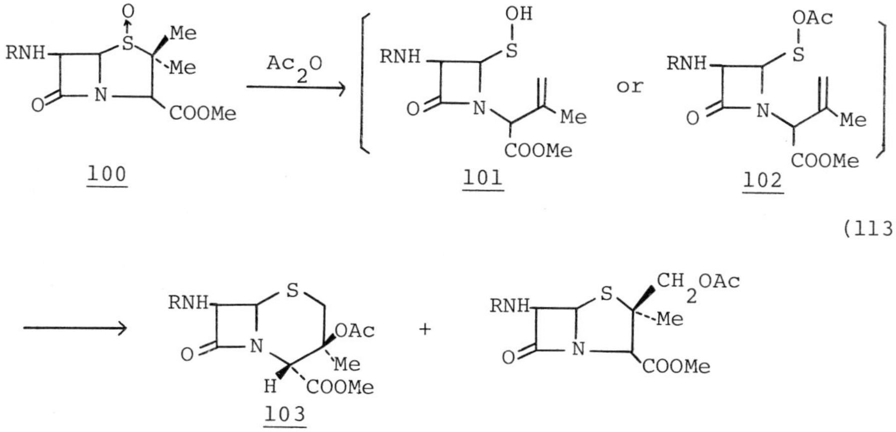

(113)

This transformation is useful for synthesis of new antibiotics,
and several recent articles have reported the successful conversions
of penicillins into cephalosporins by the treatment of penicillin
sulfoxides with electrophilic reagents such as acid anhydride,[164,165]
acids,[166] diazo compound - amine hydrochloride,[167] 2-mercaptobenzo-
thiazole followed by halogenation and DMF treatment,[168] $SOCl_2$ - Et_3N,
[169] Me_3SiCl - α-picoline[170] and N-(β-halogenoethyl)pyridinium salt
in 1,2-dihaloethane.[171]

Refluxing 2,2-dimethyl-thiochromane S-oxide with freshly distilled
acetic anhydride is also known to give the acetate in 80% yield
(eq. 114).[172]

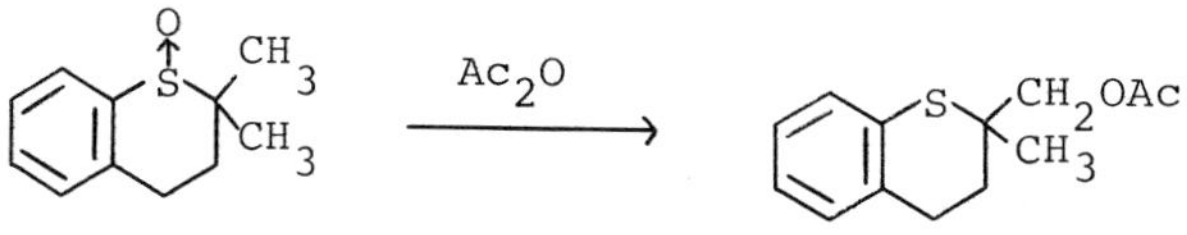

(114)

Another similar reaction, shown below (eq. 115), is also known.[173]

Apart from these reactions of penicillin or penicillin-like
sulfoxides, a ring-enlargement reaction of simple five-membered

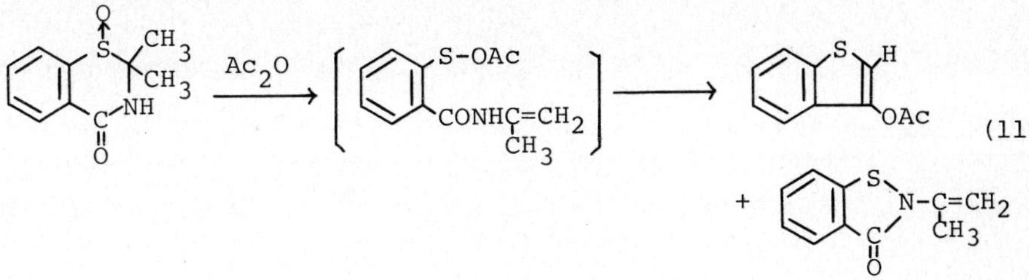

(structures for equation 115)

(115)

sulfur hetereocycles which appear to have undergone the abnormal
Pummerer reaction via the S-chloro sulfonium salt have been reported
as shown in eq. 116,[174] eq. 117,[175] and eq. 118[176].

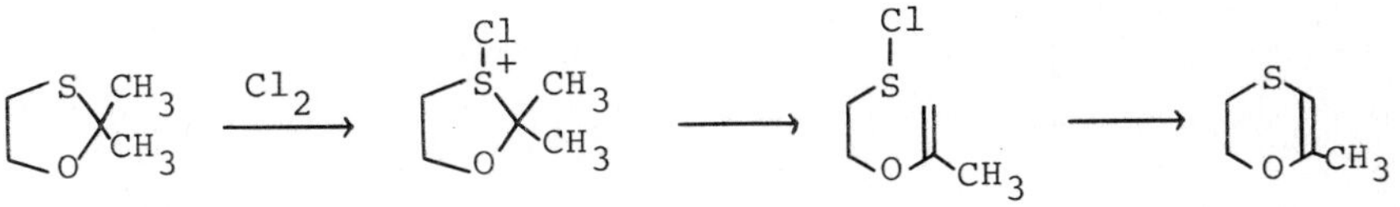

(116)

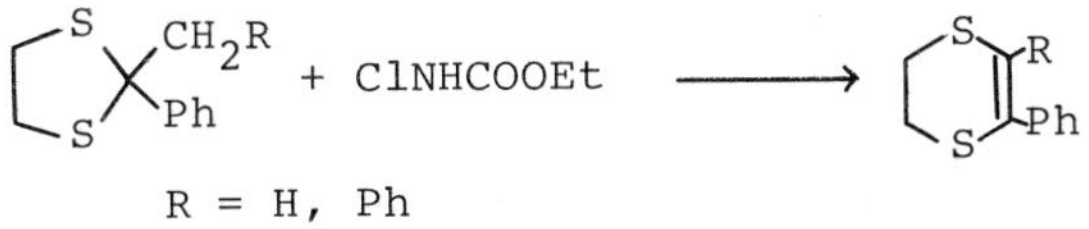

(117)

R = H, Ph

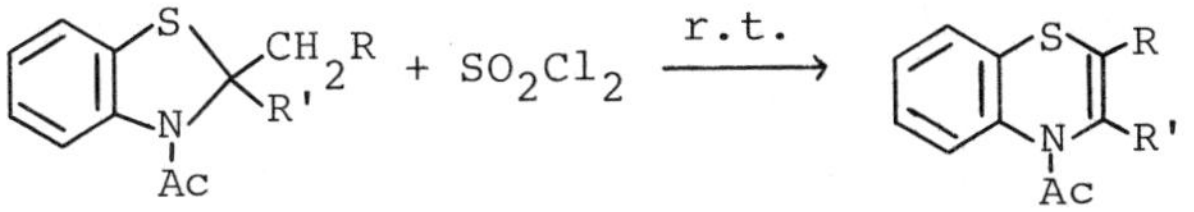

(118)

III. G. The Pummerer Reactions of Organoselenium Compounds

Organoselenium compounds show chemical behaviors analogous to
the corresponding organosulfur derivatives. When the following
selenides (104) and (105) were treated with benzoyl peroxide at
room temperature, stable "selenurane intermediates (106) and (107)"
were obtained quantitatively. These addition intermediates (106)
and (107) underwent the Pummerer rearrangement upon heating to
result in the formation of the corresponding α-benzoyloxyselenides
(eqs. 119 and 120),[32,33] as in the reaction between the sulfides
and benzoyl peroxide.[151-154]

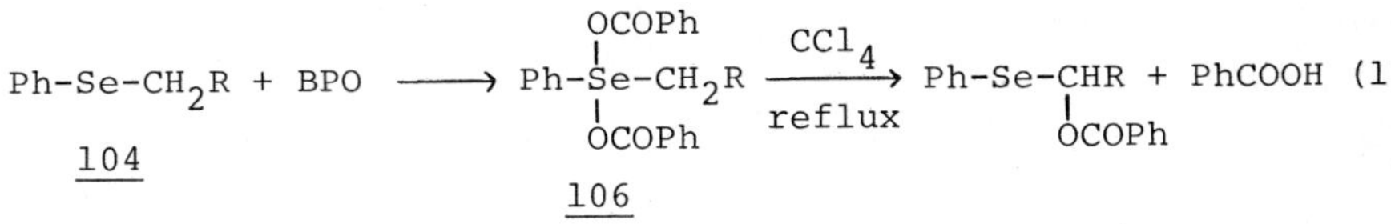

$$Ph-Se-CH_2R + BPO \longrightarrow Ph-\underset{\underset{OCOPh}{|}}{\overset{\overset{OCOPh}{|}}{Se}}-CH_2R \xrightarrow[reflux]{CCl_4} Ph-\underset{\underset{OCOPh}{|}}{Se}-CHR + PhCOOH \quad (119)$$

104   106

$$Ph-Se-CH=CH_2 \; + \; BPO \; \longrightarrow \; \underset{\underset{107}{OCOPh}}{\overset{OCOPh}{Ph-Se-CH=CH_2}} \; \overset{\Delta}{\longrightarrow} \; \underset{OCOPh}{Ph-Se-CHCH_2OCOPh} \quad (120)$$

$$\underset{105}{}$$

There are a few other examples of the Pummerer reaction of various organoselenoxides as shown in eqs. 121 and 122,[177] and eqs. 123 and 124.[178]

$$CH_3-\overset{\downarrow}{\underset{O}{Se}}-CH_3 \; + \; AcOH \; \xrightarrow{\;60°C\;} \; CH_3-Se-CH_2OAc \qquad (121)$$

$$CH_3-\overset{\downarrow}{\underset{O}{Se}}-CH_2OAc \; + \; Ac_2O \; \longrightarrow \; AcOCH_2-Se-CH_2OAc \qquad (122)$$

$$\underset{\overset{\downarrow}{O}\;\;Ph}{Ar-Se-\overset{\overset{CH_3}{|}}{\underset{|}{C}}-SiMe_3} \; \overset{\Delta}{\longrightarrow} \; \underset{OSiMe_3}{Ar-Se-\overset{\overset{CH_3}{|}}{\underset{|}{C}}-Ph} \qquad (123)$$

$$\underset{\overset{\downarrow}{O}\;\;SiMe_3}{Ar-Se-\overset{\overset{Ph}{|}}{\underset{|}{C}}-SiMe_3} \; \overset{\Delta}{\longrightarrow} \; \underset{OSiMe_3}{Ph-Se-\overset{\overset{Ph}{|}}{\underset{|}{C}}-SiMe_3} \; \longrightarrow \; \underset{O}{Ph-\overset{}{\underset{\|}{C}}-SiMe_3} \qquad (124)$$

Diphenyl teluroxide also gives the corresponding hypervalent addition compound upon treatment with acetic anhydride.[179] However, there has been no attempt to explore the reaction further with teluroxides with α-methylene substituents.

$$Ph-\overset{\downarrow}{\underset{O}{Te}}-Ph \; + \; Ac_2O \; \longrightarrow \; \underset{OAc}{\overset{OAc}{Ph-Te-Ph}}$$

III. H. Synthetic Application of the Pummerer Reaction

Since the Pummerer reaction can introduce a good electronegative functional group at α-carbon of sulfides, the reaction has versatile utility in organic syntheses of aldehydes, ketones, vinyl compounds and others, and there are numerous examples of synthetic applications. However, these applications are beyond the scope of this chapter. Therefore, only the important references are cited here: aldehydes and protected aldehydes,[46,159,180-185] α-hydroxyaldehydes,[186,187] diketones,[188,189] ninhydrine,[190,191] α-hydroxycarboxylic acid derivatives,[186,188,192] and other α-functionalized carboxylic acid

derivatives.[162,193,194] Synthesis of heterocycles by an intramolecular
Pummerer rearrangement: naphthalene derivatives,[195] indole and
carbazole derivatives,[196,197] sulfur-containing heterocycles,[25,198-200] and other heterocycles.[201-204] Olefinic sulfides have also
been synthesized, as were α- or β-mercaptovinyl ketones,[205-207]
and 1,3-diene sulfides.[202,208]

III. I. Model Reaction for Biological Demethylation

Demethylation of methionine to form homocysteine is an important
reaction in biological systems.[209] We thought that the Pummerer
reaction may serve as an enzymatic model reaction of demethylation
of methionine, and we thus treated dimethyl sulfoxide with acetyl
phenyl phosphate under moderate conditions to obtain the rearrangement
products (eq. 125).[1] We found also that methionine sulfoxide as
well undergoes demethylation by treatment with acetic anhydride to
afford eventually homocysteine in a good yield (eq. 126).[1]

$$CH_3-\underset{O}{\overset{\downarrow}{S}}-CH_3 + CH_3-\underset{O}{\overset{O}{C}}-O-\underset{OPh}{\overset{O}{P}}-ONa \xrightarrow[pH\ 5]{38°C} CH_3SCH_2OCOCH_3 \xrightarrow{H_2O} CH_3SH + CH_2O + CH_3COOH \quad (125)$$

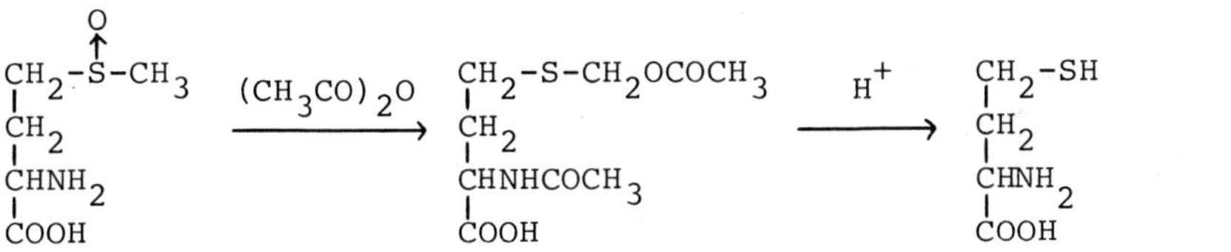

$$\begin{array}{c} CH_2-\overset{\uparrow O}{S}-CH_3 \\ CH_2 \\ CHNH_2 \\ COOH \end{array} \xrightarrow{(CH_3CO)_2O} \begin{array}{c} CH_2-S-CH_2OCOCH_3 \\ CH_2 \\ CHNHCOCH_3 \\ COOH \end{array} \xrightarrow{H^+} \begin{array}{c} CH_2-SH \\ CH_2 \\ CHNH_2 \\ COOH \end{array} \quad (126)$$

Indeed the methyl group in several methyl sulfides has been known
to be demethylated by the formation of formaldehyde in an *in vitro*
enzyme system (eq. 127).[210]

$$R-S-CH_3 + Enzyme \xrightarrow{NADPH,\ O_2} R-SH + CH_2O$$

$$(127)$$

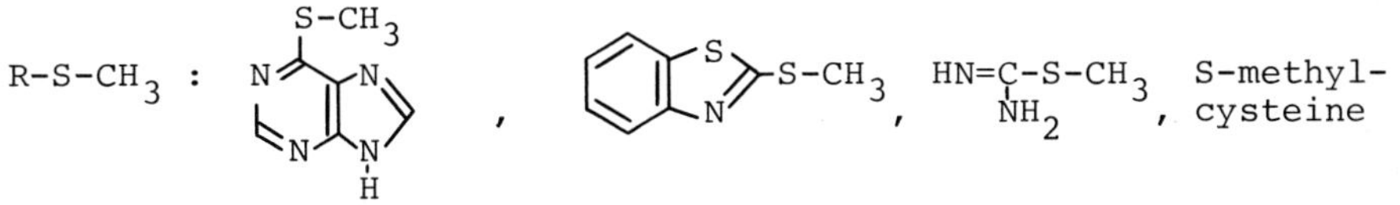

R-S-CH₃ : (structures shown)

We found recently that sulfides and disulfides are readily oxidized
with cytochrome P-450 enzyme *in vitro* to the corresponding sulfoxides

and thiolsulfinates.[211,212] Thus the demethylation of methanesulfenyl derivatives in biological systems through the Pummerer reactions may not be far from the real occurrence.

We have found recently that sulfides with electron-withdrawing substituents undergo oxidative C-S bond cleavage in the presence of a Co^{II} complex affording the corresponding disulfides and other oxidized carbonyl compounds (eq. 128).[213] The facile H-D exchange

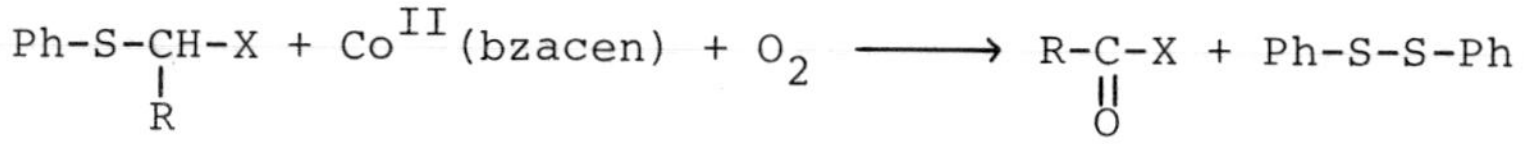

Ph-S-CH-X + Co^II(bzacen) + O2
R
R-C-X + Ph-S-S-Ph
O

(128)

X = -COPh, CN, -COOEt, $-C_6H_4-NO_2$-p

of the methylene group in the reaction system during the reaction and the sluggish nature of the reaction of the α-carbanion of the sulfide, prepared by treatment with a strong base and with O_2, and other pertinent data, seem to indicate that the reaction involves a kind of cobalt complexed sulfonium ylide intermediate (<u>108</u>) (eq. 129). This is another modification of the Pummerer reaction which could occur in biological systems.

R-S-CH2-X + Co^II(bzacen)
O2
Ph-S-CH2-X
Co^II
O2
-H+
Ph-S-CH-X
Co^II
Ph-S-CH-X
Co^II
108
Ph-S-CH-X
O-O---Co^II
X-C-H + 1/2 Ph-S-S-Ph
O

(129)

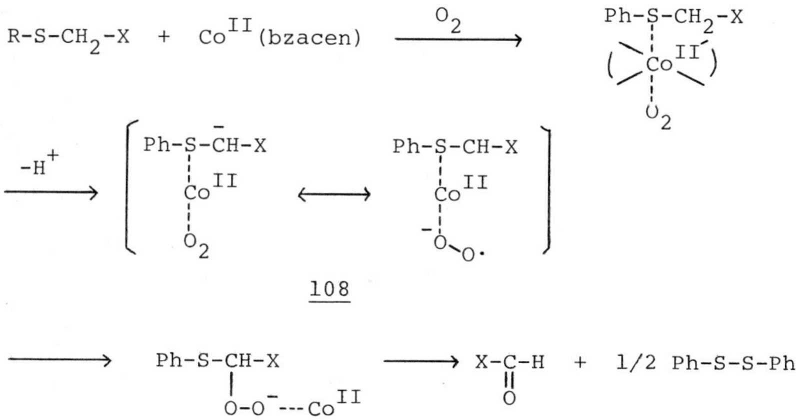

IV. REFERENCES

1 S. Oae, T. Kitao and S. Kawamura, Tetrahedron, 19(1963)1783.

2 R. Pummerer, Chem. Ber., 42(1909)2282.

3 R. Pummerer, Chem. Ber., 43(1910)1401. see also; J.A. Smythe,
J. Chem. Soc., 95(1909)349; T.P. Hilditch, Chem. Ber., 44(1911)
3583.

4 L. Horner and P. Kaiser, Ann. Chem., 626(1959)19.

5 L. Horner, Ann. Chem., 631(1960)198.

6 C.C. Price and S. Oae, Sulfur Bonding, Ronald Press, 1962, Ch. 2.

7 S. Oae, T. Kitao, S. Kawamura and Y. Kitaoka, Tetrahedron, 19
(1963)817.

8 C.R. Johnson, J.C. Sharp and W.G. Phillips, Tetrahedron Lett.,
(1967)5299.

9 C.R. Johnson and W.G. Phillips, J. Org. Chem., 32(1967)1926.

10 C.R. Johnson and W.G. Phillips, J. Am. Chem. Soc., 91(1969)682.

11 M. Murakami and S. Oae, Proc. Japan Acad., 25(1949)No. 11, 12;
J. Chem. Soc. Japan, 72(1951)595.

12 F.G. Bordwell, G.D. Cooper and H. Morita, J. Am. Chem. Soc.,
79(1957)376.

13 H. Meerwein, K.F. Zenner and R. Gipp, Ann. Chem., 688(1965)67.

14 H. Stridsberg and S. Allenmark, Acta Chem. Scand., B28(1974)591;
ibid., B30(1976)219.

15 T. Numata and S. Oae, Tetrahedron Lett., (1977)1337.

16 T. Numata and S. Oae, Yuki Gosei Kagaku Kyokaishi, 35(1977)726.

17 T. Numata, Yuki Gosei Kagaku Kyokaishi, 36(1978)845.

18 a) S. Oae in S. Oae(Ed.), Organic Chemistry of Sulfur, Plenum
Press, New York, 1977, Ch. 8.
b) S. Oae in M. Tišler(Ed.), Topics in Organic Sulfur Chemistry,
University Press, Ljubljana, Yugoslavia, 1978, Ch. 11.

19 G.A. Russell and G.J. Mikol in B.S. Thyagarajan(Ed.), Mechanisms
of Molecular Migrations, Vol. 1, Interscience, 1968, p. 157.

20 T. Durst, Advances in Organic Chemistry, 6(1969)285.

21 J.P. Marino in A. Senning(Ed.), Topics in Sulfur Chemistry,
Vol. 1, Georg Thieme Publishers, Stuttgart, 1976, p. 1.

22 A.K. Sharma and D. Swern, Tetrahedron Lett., (1974)1503.

23 S. Iwanami, S. Arita and K. Takahashi, Yuki Gosei Kagaku Kyokaishi,
26(1968)375.

24 T. Numata and S. Oae, Chem. Ind.(London), (1972)726.

25 S. Oae and T. Numata, Tetrahedron, 30(1974)2641.

26 T. Numata and S. Oae, Tetrahedron, 32(1976)2699.

27 S. Oae and M. Kise, Tetrahedron Lett., (1968)2261.

28 M. Kise and S. Oae, Bull. Chem. Soc. Japan, 43(1970)1426.

29 S. Oae and M. Kise, Tetrahedron Lett., (1967)1409.

30 S. Oae and M. Kise, Bull. Chem. Soc. Japan, 43(1970)1416.

31 M. Kise and S. Oae, Bull. Chem. Soc. Japan, 43(1970)1804.

32 Y. Okamoto, R. Homsany and T. Yano, Tetrahedron Lett., (1972) 2529.

33 Y. Okamoto, K.L. Chellapa and R. Homsany, J. Org. Chem., 38(1973) 3172.

34 S. Oae and N. Furukawa, Tetrahedron Reports No. 37, Tetrahedron, 33(1977)2359-2367.

35 P.H.W. Lau and J.C. Martin, J. Am. Chem. Soc., 99(1977)5490.

36 L.J. Kaplan and J.C. Martin, J. Am. Chem. Soc., 95(1973)793.

37 E. Vilsmaier, 7th Int'l. Symposium Org. Sulfur Chem., Hambrug (1976).

38 G.E. Wilson, Jr. and C.J. Strong, J. Org. Chem., 37(1972)2376.

39 M. Cocivera, V. Malatesta, K.W. Woo and A. Effio, J. Org. Chem., 43(1978)1140.

40 D.A. Davenport, D.B. Moss, J.E. Rhodes and J.A. Walsh, J. Org. Chem., 34(1969)3353.

41 T. Numata and S. Oae, unpublished results.

42 B. Lindberg and H. Lundström, Acta Chem. Scand., 22(1968)1861.

43 W.E. Parham and L.D. Edwards, J. Org. Chem., 33(1968)4150.

44 T. Numata, O. Itoh and S. Oae, unpublished results.

45 M. Kise and S. Oae, Bull. Chem. Soc. Japan, 43(1970)1421.

46 T.D. Harris and V. Boekelheide, J. Org. Chem., 41(1976)2770.

47 D.N. Jones, E. Helmy and R.D. Whitehouse, J. Chem. Soc., Perkin I, (1972)1329.

48 H. Kobayashi, N. Furukawa, T. Aida, K. Tsujihara and S. Oae, Tetrahedron Lett., (1971)3109.

49 N. Furukawa, T. Masuda, M. Yakushiji and S. Oae, Bull. Chem. Soc. Japan, 47(1974)2247.

50 N. Furukawa, S. Oae and T. Masuda, Chem. Ind.(London), (1975)396.

51 N. Furukawa, T. Yoshimura and S. Oae, Phosphorus and Sulfur, 3(1977)277.

52 T. Yagihara and S. Oae, Tetrahedron, 28(1972)2759.

53 S. Oae and K. Ogino, Heterocycles, 6(1977)583-675.

54 T. Numata, O. Itoh and S. Oae, Chemistry Lett., (1977)909.

55 M. Mikołajczyk, A. Zatorski, S. Grzejszczak, B. Costisella and W. Midura, J. Org. Chem., 43(1978)2518.

56 B.M. Trost and R.F. Hammen, J. Am. Chem. Soc., 95(1973)962. see also; B.M. Trost and W.G. Biddlecom, J. Org. Chem., 38(1973)3438.

57 C.A. Maryanoff, K.S. Hayes and K. Mislow, J. Am. Chem. Soc., 99(1977)4412.

58 S.J. Campbell and D. Darwish, Can. J. Chem., 54(1976)193.

59 P.K. Claus, H.A. Schwarz, W. Rieder and W. Vycudilik, Phosphorus and Sulfur, 1(1976)11.

60 T. Masuda, T. Numata, N. Furukawa and S. Oae, Chemistry Lett., (1977)903.

61 T. Masuda, T. Numata, N. Furukawa and S. Oae, J. Chem. Soc., Perkin II, (1978)1302.

62 U. Schöllkopf, E. Ruban, P. Tonne and K. Riedel, Tetrahedron Lett., (1970)5077.

63 R. Jorritsma, H. Steinberg and Th.J. de Boer, 8th Int'l. Symp. Org. Sulfur Chem., Portoroz (1978).

64 P. Calzavara, M. Cinquini, S. Colonna, R. Fornaiser and F. Montanari, J. Am. Chem. Soc., 95(1973)7431.

65 A. Garbesi and A. Fava, J. Org. Chem., 42(1977)4029.

66 S. Glue, I.T. Kay and M.R. Kipps, J. Chem. Soc., Chem. Commun., (1970)1158.

67 D.H. Bremner and M.M. Campbell, J. Chem. Soc., Chem. Commun., (1976)538; J. Chem. Soc., Perkin I, (1977)2298.

68 J.E. McCormick and R.S. McElhinney, Chem. Commun., (1969)171; J. Chem. Soc., Perkin I, (1976)2533.

69 A.G. Brook and D.G. Anderson, Can. J. Chem., 46(1968)2115.

70 F.A. Carey and O. Hernandez, J. Org. Chem., 38(1973)2670.

71 F.A. Carey, O.D. Dailey, Jr., O. Hernandez and J.R. Tucker, J. Org. Chem., 41(1976)3975.

72 E. Vedejs and M. Mullins, Tetrahedron Lett., (1975)2017.

73 E. Vilsmaier, K.H. Dittrich and W. Sprügel, Tetrahedron Lett., (1974)3601.

74 E. Vilsmaier and R. Bayer, Synthesis, (1976)46.

75 E. Vilsmaier, R. Bayer, I. Laengefelder and U. Welz, Chem. Ber., 111(1978)1136.

76 E. Vilsmaier, R. Bayer, U. Welz and K.-H. Dittrich, Chem. Ber., 111(1978)1147.

77 P.G. Gassman and R.J. Balchunis, Tetrahedron Lett., (1977)2235.

78 K.W. Ratts and A.N. Yao, J. Org. Chem., 33(1969)70.

79 A. Terada and Y. Kishida, Chem. Pharm. Bull., 18(1970)505.

80 T. Minami, Y. Tsumori, K. Yoshida and T. Agawa, J. Org. Chem., 39(1974)3412.

81 with sulfuric acid; N. Kunieda and S. Oae, Bull. Chem. Soc. Japan, 46(1973)1745.

82 with phosphoric acid; N. Kunieda and S. Oae, Bull. Chem. Soc. Japan, 41(1968)1025.

83 with chloroacetic acids; S. Oae, M. Yokoyama and M. Kise, Bull. Chem. Soc. Japan, 41(1968)1221.

84 with hydrobromic acid; W. Tagaki, K. Kikukawa, N. Kunieda and S. Oae, Bull. Chem. Soc. Japan, 39(1966)614.

85 with hydrochloric acid; a) K. Mislow, T. Simons, J.T. Melillo and A.L. Terney, Jr., J. Am. Chem. Soc., 86(1964)1452. b) H. Yoshida, T. Numata and S. Oae, Bull. Chem. Soc. Japan, 44 (1971)2875, and references in it. cf. H. Kwart and H. Omura, J. Am. Chem. Soc., 93(1971)7250.

86 R. Annunziata, M. Cinquini and S. Colonna, J. Chem. Soc., Perkin I, (1973)1231.

87 T. Masuda, T. Numata, N. Furukawa and S. Oae, Chemistry Lett., (1977)745.

88 D.K. Bates, J. Org. Chem., 42(1977)3452.

89 Y. Hiraki, M. Kamiya, R. Tanikaga, N. Ono and A. Kaji, Bull. Chem. Soc. Japan, 50(1977)447.

90 R.A. Olofson and J.P. Marino, Tetrahedron, 27(1971)4195.

91 W.E. Parham and M.D. Bharsar, J. Org. Chem., 28(1963)2686.

92 N. Kunieda, Y. Fujiwara and M. Kinoshita, 37th Annual Meeting of Japan Chemical Society, II(1978)687.

93 T. Numata, O. Itoh and S. Oae, Tetrahedron Lett., (1979)161. see also, ibid., (1979)1869.

94 G.A. Koppel and L.J. McShane, J. Am. Chem. Soc., 100(1978)288.

95 Y. Tamura, Y. Nishikawa, K. Sumoto, M. Ikeda, M. Murase and M. Kise, J. Org. Chem., 42(1977)3226.

96 D. Rittenberg and L. Ponticorvo, Int. J. Appl. Radiation and Isotopes, 1(1956)208.

97 N. Asai, Ph.D. Thesis, University of Tsukuba (1978).

98 H. Böhme, H. Fisher and R. Frank, Ann. Chem., 563(1949)54.

99 H. Böhme and H.-J. Gran, Ann. Chem., 577(1952)68.

100 G.E. Wilson, Jr. and R. Albert, J. Org. Chem., 38(1973)2160.

101 F. Boberg, Ann. Chem., 679(1964)107.

102 F. Boberg, G. Winter and G.R. Schultze, Chem. Ber., 89(1956)1160.

103 F.G. Bordwell and B.M. Pitt, J. Am. Chem. Soc., 77(1955)572.

104 W.E. Truce, G.H. Birum and E.T. McBee, J. Am. Chem. Soc., 74 (1952)3594.

105 H. Böhme and H.-J. Gran, Ann. Chem., 581(1953)133.

106 L.A. Paquette, J. Am. Chem. Soc., 86(1964)4085.

107 L.A. Paquette, L.S. Wittenbrook and K. Schreiber, J. Org. Chem.,
      33(1968)1080.

108 E. Vilsmaier and W. Sprügel, Ann. Chem., 749(1971)62.

109 E. Vilsmaier and W. Sprügel, Ann. Chem., 747(1971)151.

110 T. Masuda, N. Furukawa and S. Oae, Chemistry Lett., (1977)1103.

111 D.L. Tuleen and T.B. Stephens, Chem. Ind.(London), (1966)1555.

112 D.L. Tuleen and V.C. Marcum, J. Org. Chem., 32(1967)204.

113 D.L. Tuleen and D.N. Buchanan, J. Org. Chem., 32(1967)495.

114 D.L. Tuleen, J. Org. Chem., 32(1967)4006.

115 D.L. Tuleen and T.B. Stephens, J. Org. Chem., 34(1969)31.

116 W.H. Koster, J.E. Dolfini, B. Toeplitz and J.Z. Gougsutas,
      J. Org. Chem., 43(1978)79.

117 W. Tagaki, K. Kikukawa, K. Ando and S. Oae, Chem. Ind.(London),
      (1964)1624.

118 G. Allegra, G.E. Wilson, Jr., E. Benedetti, C. Pedone and
      R. Albert, J. Am. Chem. Soc., 92(1970)4002.

119 C.G. Kruse, N.L.J.M. Broekhof, A. Wijsman and A. van der Gen,
      Tetrahedron Lett., (1977)885.

120 A. Lüttringhaus and N. Engelhard, Chem. Ber., 93(1960)1525.

121 L.A. Paquette et al., J. Am. Chem. Soc., 93(1971)4508.

122 R.H. Rynbrandt, Tetrahedron Lett., (1971)3553.

123 E.H. Amonoo-Neizer, S.K. Ray, R.A. Shaw and B.C. Smith, J. Chem.
      Soc., (1965)6250.

124 R. Michelot and B. Tchoubar, Bull. Soc. Chim. Fr., (1966)3039.

125 R. Oda and Y. Hayashi, Tetrahedron Lett., (1967)2181, 3141.

126 M. Oki and K. Kobayashi, Bull. Chem. Soc. Japan, 43(1970)1223.

127 G.A. Russell and E.T. Sabourin, J. Org. Chem., 34(1969)2336.

128 M. Hojo and Z. Yoshida, J. Am. Chem. Soc., 90(1968)4496.

129 A. Senning, Chem. Commun., (1967)64.

130 R.E. Boyle, J. Org. Chem., 31(1966)3880.

131 C.W. Bird, J. Chem. Soc., C, (1968)1230.

132 F. Loth and A. Michaelis, Chem. Ber., 27(1894)2540.

133 M.F. Lappert and J.K. Smith, J. Chem. Soc., (1961)3224.

134 T.H. Chan, M. Milnyk and D.N. Harpp, Tetrahedron Lett., (1969)
      201.

135 K. Naumann, G. Zon and K. Mislow, J. Am. Chem. Soc., 91(1969)
      2788, 7012.

136 R. Rätz and O.J. Sweeting, Tetrahedron Lett., (1963)529; J. Org.
      Chem., 28(1963)1612.

137 A. Michaelis and B. Godchaux, Chem. Ber., 24(1891)757.

138 T. Numata and S. Oae, Chem. Ind.(London), (1973)277.

139 J. Drabowicz and S. Oae, Synthesis, (1977)404.

140 J. Drabowicz and S. Oae, Chemistry Lett., (1977)767.

141 R. Tanikaga, K. Nakayama, K. Tanaka and A. Kaji, Chemistry Lett.,
    (1977)396.

142 T. Numata, K. Ikura, Y. Shimano and S. Oae, Org. Prep. Proced.
    Int., 8(1976)119.

143 N.P. Volynskii, G.D. Gal'pern and V.V. Smolyaninov, Neftekhimiya,
    1(1961)473. Chem. Abstr., 57(1962)16510h.

144 J.S. Grossert, W.R. Handstaff and R.F. Langler, Can. J. Chem.,
    55(1977)421.

145 H. Kise, G.F. Whitfield and D. Swern, Tetrahedron Lett., (1971)
    4839.

146 I. Saito and S. Fukui, J. Vitaminol.(Kyoto), 12(1966)244. Biol.
    Abstr., 48(1967)39379.

147 K. Kondo and A. Negishi, Chemistry Lett., (1974)1525.

148 E. Block and J. O'Connor, J. Am. Chem. Soc., 96(1974)3929.

149 N. Furukawa, T. Morishita, T. Akasaka and S. Oae, Tetrahedron
    Lett., (1977)1653.

150 N. Furukawa, T. Morishita, T. Akasaka and S. Oae, Tetrahedron
    Lett., (1978)1567.

151 L. Horner and E. Jürgens, Ann. Chem., 602(1957)135.

152 a) K. Gollmer and H. Ringsdorf, Macromol. Chem., 121(1969)227.
    b) D.I. Davies, D.H. Hey and B. Summers, J. Chem. Soc., C,
       (1970)2653.

153 W.A. Pryor and H.T. Bickley, J. Org. Chem., 37(1972)2885.

154 J.E. Baldwin, A. Au, M. Christie, S.B. Harber and D. Hesson,
    J. Am. Chem. Soc., 97(1975)5957.

155 with t-butyl perbenzoate; a) G. Sosnovsky, Tetrahedron, 18(1962)
    15. b) D.J. Rawlinson and G. Sosnovsky, Synthesis, (1972)10.

156 a) D. Walker and L. Leib, Can. J. Chem., 40(1962)1242.
    b) D. Walker, J. Org. Chem., 31(1966)835.

157 H. Potter, Am. Chem. Soc. Cleaveland Meeting (1960).

158 R. Oda and K. Yamamoto, J. Org. Chem., 26(1961)4679. see also;
    P. Manya, A. Seker and R. Rumph, Tetrahedron, 26(1970)467.

159 M. Hojo, R. Masuda, T. Saeki, K. Fujimori and S. Tsutsumi,
    Tetrahedron Lett., (1977)3883.

160 S. Oae, T. Yagihara and T. Okabe, Tetrahedron, 28(1972)190.

161 L.S.S. Reasnomm and W.I. O'Sullivan, J. Chem. Soc., Chem. Commun.,
    (1976)642.

162 H. Kosugi, H. Uda and S. Yamagiwa, J. Chem. Soc., Chem. Commun.,

(1976)71.

163 R.R. King, R. Greenhalgh and W.D. Marshall, J. Org. Chem., 43 (1978)1262.

164 R.B. Morin et al., J. Am. Chem. Soc., 85(1963)1896; ibid., 91 (1969)1401.

165 D.H.R. Barton et al., J. Chem. Soc., C, (1971)3540.

166 W.J. Gottstein, P.F. Misco and L.C. Cheney, J. Org. Chem., 37 (1972)2765.

167 M. Numata, Y. Imashiro, I. Minamida and M. Yamaoka, Tetrahedron Lett., (1972)5097.

168 T. Kamiya, T. Teraji, Y. Saito, M. Hashimoto, O. Nakaguchi and T. Oku, Tetrahedron Lett., (1973)3001.

169 S. Kukolja, S.R. Lammert, M.R. Gleissner and A.I. Ellis, J. Am. Chem. Soc., 97(1975)3192.

170 J.J. de Koning, H.J. Kooreman, H.S. Tan and J. Verweij, J. Org. Chem., 40(1975)1346.

171 A. Nudelman and R.J. McCaully, J. Org. Chem., 42(1977)2887.

172 R.B. Morin, D.O. Spry and R.A. Mueller, Tetrahedron Lett., (1969)849.

173 R.B. Morin and D.O. Spry, Chem. Commun., (1970)335.

174 G.E. Wilson, Jr., J. Am. Chem. Soc., 87(1965)3785.

175 H. Yoshino, Y. Kawazoe and T. Taguchi, Synthesis, (1974)713.

176 F. Chioccara, G. Prota, R.A. Nicolaus and E. Novellino, Synthesis, (1977)876. see also; F. Chioccara, V. Mangiacapra, E. Novellino and G. Prota, J. Chem. Soc., Chem. Commun., (1977)863.

177 N. Miyoshi, S. Murai and N. Sonoda, Tetrahedron Lett., (1977) 851.

178 H.J. Reich and S.K. Shah, J. Org. Chem., 42(1977)1773.

179 S. Tamagaki, I. Hatanaka and S. Kozuka, Bull. Chem. Soc. Japan, 50(1977)2501.

180 T.L. Moore, J. Org. Chem., 32(1967)2786.

181 M. von Strandtmann, D. Connor and J. Shavel, Jr., J. Heterocyclic Chem., 9(1972)175.

182 D.T. Connor, P.A. Young and M. von Strandtmann, Synthesis, (1978)208.

183 S. Iriuchijima, T. Sakakibara and G. Tsuchihashi, Agr. Biol. Chem., 40(1976)1369.

184 Y. Nagao et al., Tetrahedron Lett., (1977)1345.

185 T.H. Chan, M. Milnyk and D.N. Harpp, Tetrahedron Lett., (1969) 201.

186 G.A. Russell and L.A. Ochrymowycz, J. Org. Chem., 34(1969)3618.

187 S. Iriuchijima, K. Maniwa and G. Tsuchihashi, J. Am. Chem. Soc.,
    96(1974)4280.

188 G.A. Russell and G.J. Mikol, J. Am. Chem. Soc., 88(1966)5498.

189 G.A. Russell and G. Hemprecht, J. Org. Chem., 35(1970)3007.

190 H.-D. Becker and G.A. Russell, J. Org. Chem., 28(1963)1896.

191 H.-D. Becker, J. Org. Chem., 29(1964)1358.

192 S. Iriuchijima, K. Maniwa and G. Tsuchihashi, J. Am. Chem. Soc.,
    97(1975)596.

193 K. Ogura and G. Tsuchihashi, J. Am. Chem. Soc., 96(1974)1960.

194 D. Hodson and G. Holt, J. Chem. Soc., C, (1968)1602.

195 Y. Oikawa and O. Yonemitsu, Chem. Commun., (1971)555; Tetrahedron
    Lett., (1972)3393; Tetrahedron, 30(1974)2653.

196 a) Y. Oikawa and O. Yonemitsu, J. Org. Chem., 41(1976)1118.
    b) Y. Oikawa and O. Yonemitsu, J. Chem. Soc., Perkin I, (1976)
       1479.

197 I. Nakagawa, H. Oka and Y. Nitta, Heterocycles, 3(1975)453.

198 D.T. Connor and M. von Strandtmann, J. Org. Chem., 39(1974)1594.

199 J. Kitchin and R.J. Stoodley, J. Chem. Soc., Chem. Commun.,
    (1972)959; J. Chem. Soc., Perkin I, (1973)22, 2464.

200 H.L. Yale, J. Heterocyclic Chem., 15(1978)331.

201 B.M. Trost and C.H. Miller, J. Am. Chem. Soc., 97(1975)7182.

202 H. Kosugi, H. Uda and S. Yamagiwa, J. Chem. Soc., Chem. Commun.,
    (1975)192.

203 A. Rosowsky and K.K.N. Chen, J. Org. Chem., 38(1973)2073.

204 A. Matsumoto, H. Shirahama, A. Ichikawa, H. Shin and S. Kagawa,
    Bull. Chem. Soc. Japan, 45(1972)1144.

205 K. Iwai, H. Kosugi and H. Uda, Chemistry Lett., (1974)1237.

206 H.J. Monteiro and A.-L. Gemal, Synthesis, (1975)437.

207 C.H. Chen, G.A. Reynolds and J.A. Van Allan, J. Org. Chem.,
    42 (1977)2777.

208 K. Praefcke and Ch. Weichsel, Tetrahedron Lett., (1976)2229.

209 G.A. Maw in A. Senning(Ed.), Sulfur in Organic and Inorganic
    Chemistry, Vol. 2, Marcel Dekker Inc., 1972, Ch. 15.

210 P. Mazel, J.F. Henderson and J. Axelroad, J. Pharmacol. Exp.
    Ther., 141(1964)1. see also; A.H. Conney, Pharmacol. Rev.,
    19(1967)317.

211 S. Oae, Y.H. Kim, D. Fukushima and T. Takata, Pure Appl. Chem.,
    49(1977)153.

212 D. Fukushima, Y.H. Kim, T. Iyanagi and S. Oae, J. Biochem.,
    83(1978)1019.

213 T. Numata, Y. Watanabe and S. Oae, Tetrahedron Lett., (1978)4933.

CHAPTER 3

APPLICATIONS OF ISOTOPIC LABELING TO THE STUDY OF
FRIEDEL-CRAFTS REACTIONS

ROYSTON M. ROBERTS AND THOMAS L. GIBSON
University of Texas at Austin

INTRODUCTION

   As in other fields of chemistry, the application of isotopic labeling to the study of Friedel-Crafts reactions
brought about numerous new and changed views concerning their
mechanisms. In addition, isotopic investigations uncovered
processes which could only become apparent in labeled systems.
   In general, a wide range of Friedel-Crafts processes have
been examined by the use of isotopic methods: (1) the interaction of a Lewis acid with an acyl or alkyl halide; (2) the
rearrangements of an alkylating species resulting from such an
interaction, as observed in the arene products or in recovered
starting materials; and (3) isomerizations of alkylation products which can occur in the presence of Lewis acids including
(a) the alkylbenzene rearrangement (internal, of side chains),
(b) disproportionation (intermolecular alkyl transfers), and
(c) reorientations, which may be the results of either intra-
or intermolecular processes.

I. Acylation

   In the first reported investigation of a Friedel-Crafts
reaction using isotopic labels, Fairbrother in 1937 carried
out the acylation of benzene with acetyl chloride in the
presence of radioactive aluminum chloride.[1] The hydrogen
chloride gas evolved by the reaction was found to contain
labeled chlorine possessing about one-fourth the original
activity found in the aluminum chloride. These results were
an elegant demonstration that all chlorine atoms of aluminum
chloride and the acetyl chloride become equivalent during
some stage in the course of the reaction. It was also found
that acetyl chloride and t-butyl chloride exchanged chloride

with labeled aluminum chloride. Fairbrother felt these data supported the concept of an ionization process of the general

$$R{-}Cl \ + \ AlCl_3 \longrightarrow \ R^+ \ + \ AlCl_4^-$$
[1]

type in the course of Friedel-Crafts reactions.

In 1961, Oulevey and Susz[2] repeated the type of experiment used by Fairbrother except that they employed benzoyl chloride rather than acetyl chloride. Complete statistical equilibration of the radioactive chloride was again observed.

II. Alkylation

A. Halide Exchange Between Alkyl and Metal Halides

In an extension of his pioneering use of isotopic labeling, Fairbrother prepared a series of radioactive metal halides.[1] He found that [82]Br-labeled aluminum bromide and stannic bromide exchange bromine readily with a variety of organic bromides at room temperature. These and other similar reports[3] gave further weight to the possibility of an ionization step in Friedel-Crafts reactions. It was also observed that the halide exchanged far more slowly in the case of a primary than tertiary alkyl halide.

A detailed investigation of the exchange of halide between an alkyl halide and aluminum halide was carried out using ethyl bromide-[82]Br.[4] The exchange reaction was studied at temperatures between 0° and -26° in various solvents. The rate of bromine exchange was extremely rapid in carbon disulfide solvent, while in nitromethane and nitrobenzene no measurable exchange was observed in the same reaction time. In order to determine if exchange could proceed via dehydrohalogenation of ethyl bromide followed by readdition of hydrogen bromide, these investigators allowed the exchange reaction to take place in the presence of deuterium bromide. As also observed in the case of n-propyl chloride with deuterium chloride, no significant incorporation of deuterium into the alkyl bromide could be detected.

In a detailed kinetic study of this halide exchange

process, third-order kinetics were found for the exchange of bromide between ethyl bromide-[82]Br and aluminum bromide in carbon disulfide solution.[5] This observation led to the conclusion that a simple ionization and recombination mechanism (eq. 2) was an inadequate explanation for the process.

$$CH_3-CH_2-Br \; + \; AlBr_3 \; \rightleftharpoons \; CH_3-CH_2^+ \; + \; AlBr_4^- \qquad [2]$$

Liquid carbon tetrachloride exchanges chlorine readily with solid aluminum chloride-[36]Cl.[6] The exchange is incomplete with short reaction times at low temperatures but is complete in 26 minutes at about 50°. In the same investigation it was noted that the n-, sec-, and tert-butyl chlorides and n-pentyl chloride exchange chloride with aluminum chloride-[36]Cl quite readily. Some isomerization apparently accompanied the halide exchange. However, while ready exchange of halide was observed with liquid carbon tetrachloride in contact with solid aluminum chloride, no detectable exchange of halide was observed to occur when both the labeled aluminum chloride and carbon tetrachloride were allowed to mingle in the vapor phase at temperatures up to 140°.

Although the observed exchange of halide between catalyst and alkylating agent seemed to support the proposed ionization step in the mechanism for Friedel-Crafts alkylation, there was no clear evidence of the nature of the ionized alkylating species which might be regarded as a fully developed cation or other related positively charged fragment.

B. Molecular Rearrangements in Alkyl Halides and Alkanes

1. Ethyl Halides. Another advance in the investigation of Friedel-Crafts alkylation by use of isotopic labeling involved a study of the alkylation of benzene with ethyl-2-^{14}C chloride in the presence of aluminum chloride.[7] The product of the reaction at 80° was ethyl-β-^{14}C-benzene; thus no rearrangement of the labeled ethyl group accompanied the alkylation. If, however, the starting labeled ethyl chloride was allowed to stand in contact with aluminum chloride for one hour at room temperature before addition of benzene, the alkylation product

was ethylbenzene in which there was essentially equal distribution of ^{14}C between the α- and β-positions in the side
chain. Thus, although labeled ethyl chloride is isotopically
isomerized by aluminum chloride in the absence of benzene,
under the usual alkylation conditions the alkylation of
benzene is much faster than the isomerization.

More detailed investigation of a similar isotopic rearrangement was conducted using ethyl-1-^{14}C bromide and aluminum
bromide.[4] A kinetic study of this reaction led to the assign-
ment of a value of approximately 19.1 kcal/mole for the energy
of activation of the isotopic isomerization in ethyl bromide.

The isomerization accompanying Friedel-Crafts alkylation of
benzene by ethyl-2-^{14}C-bromide[4] and by ethyl-2-^{14}C iodide[8] in
the presence of aluminum halides was also studied. Whereas
ethyl-2-^{14}C chloride underwent almost complete equilibration
of the isotopic label during one hour in the presence of
aluminum chloride at room temperature, complete equilibration
in labeled ethyl bromide by aluminum bromide was only achieved
during eight hours at 80°,[4b] and in labeled ethyl iodide by
aluminum chloride after about forty-two hours at 40°.[8] Among
ethyl halides, the observed order of reactivity, RCl > RBr >
RI, is not surprising and may simply be rationalized in terms
of the order of polarization of the carbon halogen bond.

More recently Nakane, Kurihara, and Natsubori[9] reported
that in the alkylation of benzene and toluene using a mixture
of ethyl-2-^{14}C iodide and aluminum chloride in n-hexane, 20%
of the ^{14}C-label was rearranged in the side chain of the
ethylbenzene or ethyltoluene produced. Similarly, when
benzene was alkylated with ^{14}C-labeled ethyl fluoride in the
presence of boron trifluoride in n-hexane solution, with a
10/1 solvent/benzene mole ratio and reaction time of 120
hours, 48% rearrangement of the label was observed.[10] Under
similar conditions using a reaction time of 17 hours and
cyclohexane as solvent, 34% of the ^{14}C-label was rearranged.
Rearrangement after one hour with the same conditions except
with nitromethane as solvent was barely measureable. In
Nakane's very dilute solutions, the reaction with benzene in
n-hexane or cyclohexane was slowed and did not compete

well with the intramolecular rearrangement of ^{14}C-label in ethyl iodide and ethyl fluoride. Only a small amount of alkylation product, in approximately 5% yield, was formed in each case under these unfavorable conditions.

Natsubori and Nakane[10] also reported that competitive alkylations of toluene and benzene by ethyl fluoride gave anomalous values of k_T/k_B = 0.56 in hexane solution and 0.57 in cyclohexane solution. These data were criticized by DeHaan and co-workers,[11] who showed that they were probably the result of phase separation, and that in truly homogenous solution the normal ratio of k_T/k_B was >1 and a relatively high percentage of meta-ethylation occurred, in agreement with Brown's Selectivity Relationship.[12]

2. Phenylethyl Halides. In the alkylation of anisole with 2-phenylethyl-1-^{14}C chloride or 2-phenylethanol-1-^{14}C in the presence of aluminum chloride in a large excess of anisole as solvent Lee, Foreman, and Rosenthal[13] reported complete equilibration of the label between the two alkyl carbons of the product, 1-p-anisyl-2-phenylethane-^{14}C, which was obtained in only 10% yield (eq. 3). None of the starting chloride and no 1-p-anisyl-1-phenylethane-^{14}C could be isolated.

$$C_6H_5CH_2{}^{14}CH_2Cl \ + \ C_6H_5OCH_3$$

$$AlCl_3 \Big\downarrow C_6H_5OCH_3$$

$$\underline{p}\text{-}CH_3OC_6H_4{}^{14}CH_2CH_2C_6H_5 \ + \ \underline{p}\text{-}CH_3OC_6H_4CH_2{}^{14}CH_2C_6H_5 \qquad [3]$$

      not rearranged   rearranged

Lee et al. proposed a 1,2-phenyl shift preceding or accompanying the alkylation or intervention of a carbocation intermediate to explain the scrambling of the label. Deactivation of the aluminum chloride catalyst through formation of a complex with anisole may account for the low yield. The rearrangement might then be related to the slowness of the alkylation reaction. Lee et al. pointed out that the

rearrangement may arise during a separate process prior to the
alkylation step, the rate of the catalyst-promoted rearrangement being greater or equal to the rate of alkylation.
Plainly, alkylation of toluene under the same conditions and
recovery of unreacted starting 2-phenylethyl chloride could
provide a useful check of these results.

In 1964, McMahon and Bunce[14] studied the same rearrangement
of 2-phenylethyl chloride with toluene as the substrate
(Scheme 1). The alkylation of toluene with 2-phenylethyl-1-^{14}C
chloride in the presence of aluminum chloride gave 1-p-tolyl-
2-phenylethane in which the ^{14}C-label was distributed 52% in
the 2-position and 48% in the 1-position (eq. 6). The authors
ascribed the apparent small difference in label distribution
between the carbons to a ^{14}C-isotope effect. Labeled 2-
phenylethyl chloride was recovered after the alkylation experiment and showed no label rearrangement.

These findings, like those of Lee, Foreman, and Rosenthal,[13]
supported a mechanism involving a 1,2-phenyl shift or formation of a phenonium ion intermediate (2) in an irreversible
reaction step preceding the alkylation (eq. 5).

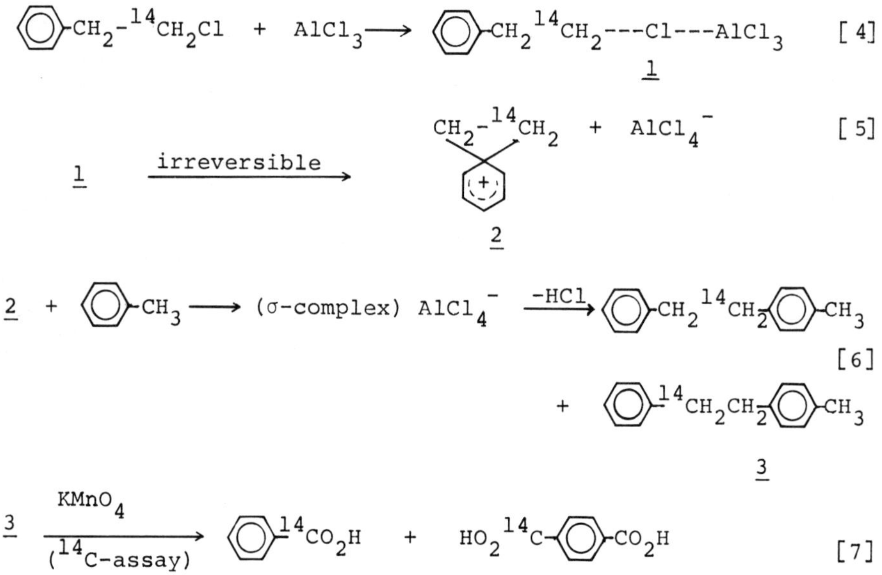

$$\text{C}_6\text{H}_5\text{-CH}_2\text{-}^{14}\text{CH}_2\text{Cl} + \text{AlCl}_3 \longrightarrow \text{C}_6\text{H}_5\text{-CH}_2\text{-}^{14}\text{CH}_2\text{---Cl---AlCl}_3 \qquad [4]$$

1

$$1 \xrightarrow{\text{irreversible}} \quad \underset{2}{\overset{\text{CH}_2\text{-}^{14}\text{CH}_2}{\bigtriangleup_{(+)}}} + \text{AlCl}_4^- \qquad [5]$$

$$2 + \text{C}_6\text{H}_5\text{-CH}_3 \longrightarrow (\sigma\text{-complex}) \text{ AlCl}_4^- \xrightarrow{-\text{HCl}} \text{C}_6\text{H}_5\text{-CH}_2\text{-}^{14}\text{CH}_2\text{-C}_6\text{H}_4\text{-CH}_3 \qquad [6]$$

$$+ \quad \text{C}_6\text{H}_5\text{-}^{14}\text{CH}_2\text{CH}_2\text{-C}_6\text{H}_4\text{-CH}_3$$

3

$$3 \xrightarrow[(^{14}\text{C-assay})]{\text{KMnO}_4} \text{C}_6\text{H}_5\text{-}^{14}\text{CO}_2\text{H} + \text{HO}_2\text{-}^{14}\text{C-C}_6\text{H}_4\text{-CO}_2\text{H} \qquad [7]$$

SCHEME 1

The alkylation of toluene under the same conditions using 2-p-nitrophenylethyl-1-^{14}C chloride produced the corresponding alkylation product but showed only 8% isotopic rearrangement. The lack of greater rearrangement in this case was explained as a result of nucleophilic attack by toluene being favored over formation of a phenonium ion intermediate which would be destabilized by the presence of the nitro group (Scheme 2).[14]

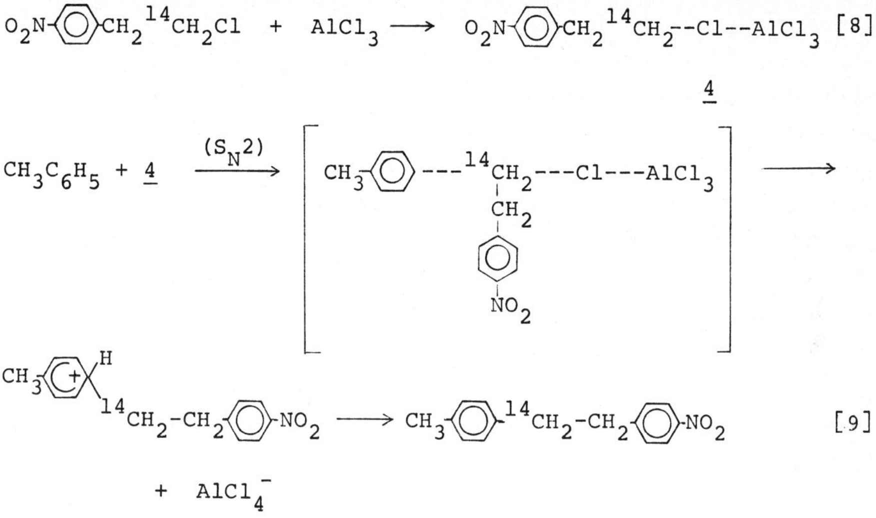

$$O_2N-\langle \bigcirc \rangle-CH_2\,^{14}CH_2Cl \ + \ AlCl_3 \longrightarrow \ O_2N-\langle \bigcirc \rangle-CH_2\,^{14}CH_2--Cl--AlCl_3 \quad [8]$$

4

$$CH_3C_6H_5 + \underline{4} \xrightarrow{(S_N2)} \left[CH_3-\langle \bigcirc \rangle ---\,^{14}CH_2---Cl---AlCl_3 \right] \longrightarrow$$

with CH$_2$, phenyl ring bearing NO$_2$

$$CH_3-\langle \overset{H}{\underset{}{C^+}} \rangle \atop \,^{14}CH_2-CH_2-\langle \bigcirc \rangle-NO_2 \longrightarrow CH_3-\langle \bigcirc \rangle-\,^{14}CH_2-CH_2-\langle \bigcirc \rangle-NO_2 \quad [9]$$

$$+ \ AlCl_4^-$$

SCHEME 2

In a recent publication, Olah and co-workers[15] described studies of methyl and ethyl fluoroantimonate complexes by methods including isotopic labeling with deuterium and ^{13}C. Intramolecular scrambling of label in the ethyl complex and even intermolecular exchange of deuterium were observed under these strongly acidic conditions.

3. Propyl Halides. The interaction of n-propyl chloride and aluminum chloride results in isomerization to isopropyl chloride. A possible path for this isomerization could be dehydrohalogenation to propylene followed by readdition of hydrogen chloride to form isopropyl chloride. Many of the halogen exchange reactions of alkyl halides with aluminum halides mentioned earlier could possibly proceed via a similar

path. A test of this possibility was conducted by allowing n-propyl chloride to isomerize in the presence of deuterium chloride and aluminum chloride at $0°$.[16] The isopropyl chloride produced was analyzed for deuterium content by mass spectrometry and was found to be devoid of deuterium. Under essentially identical conditions propene readily added deuterium chloride in the presence of aluminum chloride to form deuterioisopropyl chloride. These data quite clearly demonstrated that a mechanism such as that of eq. 10, involving dehydrohalogenation and readdition of hydrogen halide, could not play any significant part in the processes concerned. An alternative route (eq. 11), which involves catalyst-induced

$$CH_3-CH_2-CH_2-Cl \longrightarrow CH_3-CH=CH_2 + HCl \longrightarrow CH_3-\overset{\overset{\displaystyle Cl}{|}}{CH}-CH_3 \quad [10]$$

$$CH_3-CH_2-CH_2-Cl \xrightarrow{Cl^-} CH_3-CH_2-\overset{+}{CH_2} \xrightarrow{\sim H:^-}$$

$$CH_3-\overset{+}{CH}-CH_3 \xrightarrow{+Cl^-} CH_3-\overset{\overset{\displaystyle Cl}{|}}{CH}-CH_3 \quad [11]$$

ionization of the n-propyl chloride followed by a 1,2-hydrogen shift and recombination with halide ion, was suggested.

In 1964, Douwes and Kooyman[17] performed aluminum bromide-catalyzed isomerization of n propyl bromide in the presence of deuterium bromide, and again, in agreement with the above results, no incorporation of deuterium was observed, indicating that the isomerization did not occur via a reversible dehydrohalogenation. The authors also treated 1-bromopropane-2-d and 1-bromopropane-2,2-d_2 with aluminum bromide and found a significant isotope effect, with hydrogen migrating 3.4 times faster than deuterium. On the basis of the above facts and finding that the exchange between organic and inorganic bromine was faster than isomerization to isopropyl bromide, Douwes and Kooyman proposed that the mechanism involved a fast reversible formation of an intermediate in which the bromine atoms have become scrambled ("ionization step") (eq. 12) followed by a slow hydrogen migration within this intermediate leading to isomerization (eq. 13).

$$\underline{n}\text{—AlBr}_3 \;+\; \underline{n}\text{—C}_3\text{H}_7\text{Br} \;\xrightleftharpoons{\text{fast}}\; (\underline{n}\text{—C}_3\text{H}_7{}^+\text{Al}_n\text{Br}_{3n+1}{}^-) \qquad [12]$$

$$(\underline{n}\text{—C}_3\text{H}_7{}^+\text{Al}_n\text{Br}_{3n+1}{}^-) \;\xrightleftharpoons{\text{slow}}\; \text{CH}_3\text{CHBrCH}_3 \qquad [13]$$

It was pointed out that in the above equations, the $\underline{n}$-propyl cation should not be regarded as a fully developed cation but as a $\underline{n}$-propyl fragment carrying a significantly greater positive charge than in free $\underline{n}$-propyl bromide. Karabatsos and co-workers[18, 19] conducted further studies of the interconversion of $\underline{n}$-propyl bromide and isopropyl bromide by treating 1-bromopropane-1,1-$\underline{d}_2$, -2,2-$\underline{d}_2$, and -1-^{13}C as well as 2-bromopropane-2-$\underline{d}$ and -1,1,1,3,3,3-$\underline{d}_6$ with aluminum bromide at $0°$. Treatment of either 1-bromopropane or 2-bromopropane gave an equilibrium mixture of about 6% 1-bromopropane and 94% 2-bromopropane under these conditions. An isotope effect (k_H/k_D) on the 1,2-hydride shift leading to 2-bromopropane was observed in these experiments in agreement with the one reported by other workers.[17] Analysis of 2-bromopropane produced during partial isomerization of 1-bromopropane-1,1-$\underline{d}_2$ and 1-bromopropane-1-^{13}C in the presence of aluminum bromide by means of nuclear magnetic resonance and mass spectrometric techniques gave the following data summarized under equations 14 and 15 (adjusted to 100% ^{13}C-labeling).

$$\text{CH}_3\text{CH}_2\text{CD}_2\text{Br} \longrightarrow \underset{\text{Br}}{\text{CH}_3\text{CHCHD}_2} \;+\; \underset{\text{Br}}{\text{CH}_3\text{CDCH}_2\text{D}} \;+\; \underset{\text{Br}}{\text{CH}_2\text{DCHCH}_2\text{D}} \quad [14]$$

100 percent $\underline{d}_2$

45% conversion	98.7%	1.3%
79% conversion	98.0%	2.0%
80% conversion	97.0%	3.0%

$$\text{CH}_3\text{CH}_2{}^{13}\text{CH}_2\text{Br} \longrightarrow \underset{\text{Br}}{\text{CH}_3\text{CH}^{13}\text{CH}_3} \;+\; \underset{\text{Br}}{\text{CH}_3{}^{13}\text{CHCH}_3} \qquad [15]$$

99.5%	0.5%

Thus, the 2-bromopropanes were mainly isotope-position unrearranged and must be produced from an essentially irreversible
1,2-hydride shift.

Similar data about n-propyl to isopropyl rearrangements
were obtained by Lee and Woodcock[20] from the alkylation of
benzene by 1-chloropropane-1,1-d$_2$ in the presence of aluminum
chloride.

In a later publication, Lee and Woodcock[21] reported investigation of isotopic rearrangement during partial isomerization of 1-^{14}C-1-chloropropane to 2-chloropropane upon treatment with aluminum chloride at 0°. They found that the
extents of isotopic rearrangement and n-propyl to isopropyl
isomerization could be influenced by changes in the reaction
conditions, and that a small amount of ^{14}C activity (1.4 to
3.0%) rearranged to the C-2 positions of the product, 2-chloro-
propane. This result was in contrast to that reported by
Karabatsos, et al.,[19] who found that in the partial isomerization of 1-bromopropane-1-^{13}C in the presence of aluminum
bromide, the resulting 2-bromopropane had essentially no label
rearranged to the C-2 positions. However, the apparent difference between these results of Karabatsos et al.,[19] and
those of Lee et al.,[21] may be considered negligible as noted
by the latter authors, and it may be concluded that only very
slight isotopic rearrangement in the 2-halopropane could have
occurred in either of the two sets of experiments. Lee et al.
also mentioned that Deno, in private communications, pointed
out that, during such isomerizations of labeled n-propyl
halide to isopropyl halide, some isotope position rearrangements can occur in the n-propyl halide. To explain these observations, Lee and Woodcock[21] proposed that the mechanism for
the conversion of n-propyl chloride to isopropyl chloride involved formation of a propyl cation-AlCl$_4$ anion ion pair or
complex in which the positive moiety could behave as a classical carbonium ion and could, among other things, undergo an
irreversible 1,2-hydride shift to give an isopropyl cation.

Isotopic n-Propyl Rearrangements. Protonated Cyclopropane
Intermediates. Following a report by J. D. Roberts and

M. Halman[22] in 1953 that deamination of 1-aminopropane-1-^{14}C
in 35% perchloric acid solution gave rise to 1-propanol-^{14}C in
which some of the ^{14}C-label had rearranged to the C-2 position, the deamination of labeled 1-aminopropane and solvolysis
of labeled n-propyl tosylates were extensively studied. Isotopic rearrangement of label from C-1 to C-2 and C-3 was detected in several cases. Mechanisms involving protonated
cyclopropane intermediates were developed to explain the various isotopic rearrangements.[18, 19, 23-31]

When isotopically labeled n-propyl halides were treated
with Friedel-Crafts catalysts, additional results with a
bearing on the intermediacy of protonated cyclopropanes were
reported. In 1963, Reutov and Shatkina[32, 33] reported that
treatment of 1-chloropropane-1-^{14}C with $ZnCl_2$-HCl produced
partial rearrangement to 1-chloropropane-3-^{14}C (eq. 17). Interestingly, no interconversion between 1- and 2-chloropropane
was noted under the applied conditions. When the reaction was
effected in the presence of aluminum chloride, however, rapid
formation of isopropyl chloride was observed (eq. 16). The
authors explained their results by suggesting that in the
reaction with zinc chloride-hydrochloric acid a 1,3-migration
of the hydride ion occurs within a strongly polarized molecule. Susan also reported observing a 1,3-hydride shift in

$$CH_3CH_2{}^{14}CH_2Cl \xrightarrow[\text{very rapid}]{AlCl_3} CH_3\underset{\underset{Cl}{|}}{CH}{}^{14}CH_3 \qquad [16]$$

$$CH_3CH_2{}^{14}CH_2Cl \xrightarrow{ZnCl_2, HCl} {}^{14}CH_3CH_2CH_2Cl \qquad [17]$$

the reaction of 1-bromopropane-1-d upon treatment with $ZnCl_2$-
HCl to yield 1-bromopropane-3-d.[34]

More recently several attempts were made to verify this
reported 1,3-hydride shift. In 1969, Lee and co-workers[35]
studied the product of treatment of 1-chloropropane-1-t with
$ZnCl_2$-HCl under reflux at 50 ± 2° for 100 hr. The recovered
1-chloropropane was uncontaminated by 2-chloropropane as reported by Reutov and Shatkina, but it showed no rearrangement
of the t-label to either C-2 or C-3 in contrast to the earlier

report.

In 1970, Karabatsos et al.[36] discussed the question of protonated cyclopropanes in the reaction of 1-chloropropane with
$ZnCl_2$-HCl. When 1-chloropropane-1,1-d_2 and -2,2-d_2 were
treated with $ZnCl_2$-HCl at $50°$ for 72 hr., the recovered products were shown to consist only of isotope-position unrearranged 1-chloropropanes; glpc analysis indicated less than
0.2% 2-chloropropane. These results, similar to those obtained by Lee et al.[35] with 1-chloropropane-1-t, failed to
confirm the intervention of either protonated cyclopropanes or
the 1,3-hydride shift reported by Reutov et al.[28, 29, 32, 33]
and Susan[34] in the reaction of 1-halopropanes with $ZnCl_2$-HCl.

Previously, it was pointed out that Karabatsos et al.[19]
studied the reaction of isotopically labeled 1-bromopropane in
the presence of aluminum bromide. Careful analysis by nmr and
mass spectrometry of the 1-bromopropane recovered after partial conversion to 2-bromopropane indicated the isotopic distributions shown in equations 18 to 20.

$$CH_3CH_2CD_2Br \xrightarrow[\text{conversion}]{80\%} C_2H_5CD_2Br + C_2H_4DCHDBr + C_2H_3D_2\text{-}CH_2Br$$

100% d_2 79.8%   5.0%   15.2% [18]

$$CH_3CD_2CH_2Br \xrightarrow[\text{conversion}]{65\%} 2.1\% \qquad\qquad 5.8\% \qquad\qquad 92.1\% \ [19]$$

100% d_2

$$CH_3CH_2{}^{13}CH_2Br \xrightarrow[\text{conversion}]{80\%} CH_3CH_2{}^{13}CH_2Br + CH_3{}^{13}CH_2CH_2Br$$

100% ^{13}C 85.7 ± 0.2%   3.7 ± 0.9%

$$+ \quad {}^{13}CH_3CH_2CH_2Br$$

10.6 ± 0.6%   [20]

In rationalizing their results, the authors concluded that
the formation of isotope-position rearranged 1-bromopropanes
could best be explained in terms of a protonated cyclopropane
mechanism. In that mechanism, a single process involving
equilibration between edge-protonated cyclopropane inter-

mediates, as pictured in Scheme 3, was believed responsible
for scrambling both the carbons and the hydrogens of the n-
propyl system.

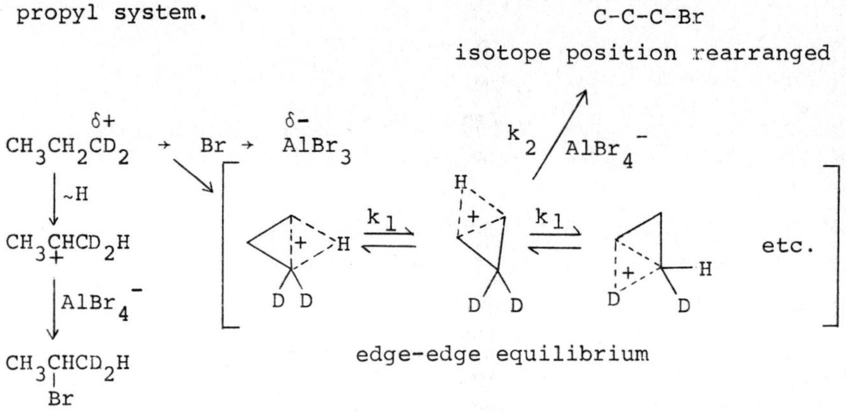

C-C-C-Br

isotope position rearranged

Scheme 3

It was emphasized that for reactions proceeding by such a
mechanism, various degrees of scrambling would be observed,
depending on the relative rates of edge-edge equilibration
(k_1) and nucleophilic attack on the protonated cyclopropane
(k_2). If $k_1 \gg k_2$, complete equilibration of all of the three
carbons would occur; on the other hand, if $k_2 \gg k_1$, the mechanism would correspond to a nominal 1,3-hydride shift and only
scrambling of C-1 and C-3 would take place. In the aqueous
deamination of 1-aminopropane the ratio of k_1/k_2 was estimated
to be between five and ten.[26, 37] In the present reaction of
1-bromopropane with aluminum bromide, this ratio was thought
to be somewhat smaller than that of deamination, but still
greater than one.

Going back to equation 20, it is important to note that,
after the short reaction times (5-6 minutes) leading to these
results, about one-third of the rearranged ^{13}C label had mi-
grated to C-2 and two-thirds to C-3. This particular finding
served to rule out any significant contribution from the
methyl-bridged intermediate (a) to the isotope-position rearrangement of the recovered 1-bromopropane. Such a species
would lead to at least as much ^{13}C at C-2 as at C-3 of the
propyl system. To account for the unequal distribution of
activity between C-2 and C-3, it was assumed that more product

$$
\begin{array}{c}
CH_3 \\
\diagup\ +\ \diagdown \\
CH_2 \overline{}^{13}CH_2
\end{array}
$$

(a)

is formed from the initially formed edge protonated cyclopropane than from the other isotope-position isomers potentially
in equilibrium with it.[37]

   In the same investigation, Karabatsos et al.[19] investigated
the treatment of isotopically labeled 1-bromopropanes with
aluminum bromide for long reaction times. Under such conditions more extensive rearrangements were observed. For example, as shown by equation 21, the 1-bromopropane-^{13}C recovered after treating 1-bromopropane-1-^{13}C for 180 minutes
showed that 52% of the isotopic label now resided at carbons
2 and 3. The lack of statistical scrambling, even after such

$$
CH_3CH_2{}^{13}CH_2Br \xrightarrow[\text{AlBr}_3]{\text{180 min.}} CH_3CH_2{}^{13}CH_2Br + \left[\begin{array}{c} {}^{13}CH_3CH_2CH_2Br \\ + \\ CH_3{}^{13}CH_2CH_2Br \end{array}\right] \quad [21]
$$

$$
 48\% 52\%
$$

prolonged reaction time, was attributed by the authors to deactivation of the catalyst by polymeric material formed during
the reaction.

   In connection with the results of Karabatsos et al.[19] it is
important to note that these authors excluded the bimolecular
reactions shown in equation 22 as mechanistic paths responsible for the isotope-position rearranged products, as no
double labeled carbon-13 species were detected, even after

$$
\overset{*}{C}-C = C + C-\overset{+}{\underset{}{C}}-\overset{*}{C} \rightleftharpoons \overset{*}{C}-\overset{+}{\underset{}{C}}-C-\overset{\overset{\displaystyle C}{|}}{C}-C \rightleftharpoons \rightleftharpoons
$$

$$
\overset{*}{C}-\overset{\overset{\displaystyle *C}{|}}{C}-C-C-C \underset{\longleftarrow}{\longrightarrow} \overset{*}{C}-C-\overset{*}{C} + C-C=C \quad [22]
$$

$$
 + +
$$

reaction for 180 minutes. (Such bimolecular reactions were
shown by Karabatsos and his coworkers to constitute a significant pathway in the isotope-position rearrangement of t-

pentyl chloride with aluminum chloride; this is described in a later section.)

More recently, Lee and Woodcock[21] reported studies of the aluminum chloride-induced partial isomerization of 1-chloro-propane-1-^{14}C to isotopically scrambled 1-chloropropane and 2-chloropropane at 0°. The extent of isotopic scrambling from C-1 to C-2 and C-3 in the recovered 1-chloropropane was determined by degradation through conversions to propionic acid, to acetic acid, and to methylamine. The authors reported that when the isomerization of 1-chloropropane to 2-chloropropane was 90 percent complete, the recovered 1-chloropropane-^{14}C showed about 7% and 22% rearrangements of the label from C-1 to C-2 and C-3, respectively. To account for the scrambling of label to C-2, these authors proposed the intervention of equilibrating protonated cyclopropane intermediates which would collapse to 1-chloropropanes with the ^{14}C-label rearranging to both C-2 and C-3. Lee and Woodcock[21] also commented on the fact that, in these reactions, more of the rearranged isotope was found at C-3 than C-2. They pointed out that this finding could be reasonably explained in terms of two factors: a) the involvement of edge-protonated cyclopropane structures, and b) the reversible isomerizations between 1- and 2-chloropropane as depicted in equation 23. Apparently, a combination of both factors is responsible for locating more of the rearranged label at C-3 than C-2.

$$CH_3CH_2{}^{14}CH_2Cl \rightleftharpoons CH_3CHCl{}^{14}CH_3 \rightleftharpoons ClCH_2CH_2{}^{14}CH_3 \qquad [23]$$

In connection with the AlX_3-induced isomerizations of iso-topically-labeled n-propyl halides, it is interesting to note that in spite of all of the information available about the nature of the isomerized products, uncertainty still prevails as to their exact mode of formation.[31]

4. t-Pentyl Chloride. J. D. Roberts, McMahon, and Hine[38] observed that ^{14}C-label was scrambled among the carbons of tert-pentyl chloride recovered after treatment with aluminum chloride (Scheme 4). However, since the rate of the process

118

shown in eq. 24 was not twice the rate of that in eq. 25 as
expected if eq. 26 were the sole pathway of isotopic rearrangement, a mechanism such as that of eq. 27 involving primary carbocations was thought to accompany the pathway of eq.
26.

$$
C-{}^{14}\overset{\displaystyle C}{\underset{\displaystyle Cl}{C}}-C-C
\;\xrightleftharpoons{AlCl_3}\;
C-\overset{\displaystyle C}{\underset{\displaystyle Cl}{C}}-{}^{14}C-C
\qquad [24]
$$

$$
{}^{14}C-\overset{\displaystyle C}{\underset{\displaystyle Cl}{C}}-C-C
\;\xrightleftharpoons{AlCl_3}\;
C-\overset{\displaystyle C}{\underset{\displaystyle Cl}{C}}-C-{}^{14}C
\qquad [25]
$$

$$
C-\overset{\displaystyle }{\underset{\displaystyle +}{{}^{14}C}}-C-C
\;\underset{\sim H}{\rightleftharpoons}\;
C-{}^{14}\overset{\displaystyle C}{\underset{\displaystyle +}{C}}-C-C
\;\overset{\sim Me}{\rightleftharpoons}\;
C-{}^{14}\overset{\displaystyle C}{\underset{\displaystyle +}{C}}-C-C
\;\overset{\sim H}{\rightleftharpoons}\;
C-{}^{14}\overset{\displaystyle C}{C}-\overset{}{C}-C
\quad [26]
$$

$$
{}^{14}C-\overset{\displaystyle C}{\underset{\displaystyle +}{C}}-C-C
\;\underset{\sim Me}{\rightleftharpoons}\;
{}^{14}C-\overset{\displaystyle C}{\underset{\displaystyle C}{C}}-C^{+}
\;\overset{\sim {}^{14}Me}{\rightleftharpoons}\;
C-\overset{\displaystyle C}{\underset{\displaystyle +}{C}}-C-{}^{14}C
\qquad [27]
$$

Scheme 4

Further evidence concerning the rearrangements of _tert_-
pentyl carbocations was obtained by Karabatsos, Vane, and
Meyerson[39] by studying the reactions of $1-{}^{13}C$ and $2-{}^{13}C-$_tert_-
pentyl chlorides (43.3% unlabeled and 57.7% monolabeled molecules). Labeled _tert_-pentyl chloride recovered after treatment with a small amount of aluminum chloride (approximately a
33:1 mole ratio) for five minutes at $0°C$ was subjected to mass
spectrometry and proton nmr analysis. It was learned that
4.3% of molecules of _tert_-pentyl chloride after treatment contained two ${}^{13}C$ atoms. This finding clearly showed that bimolecular reactions involving C_{10}-carbocation   together
with the unimolecular methyl shifts, are responsible for the
isotope-position rearrangements in the _tert_-pentyl system
(Scheme 5).

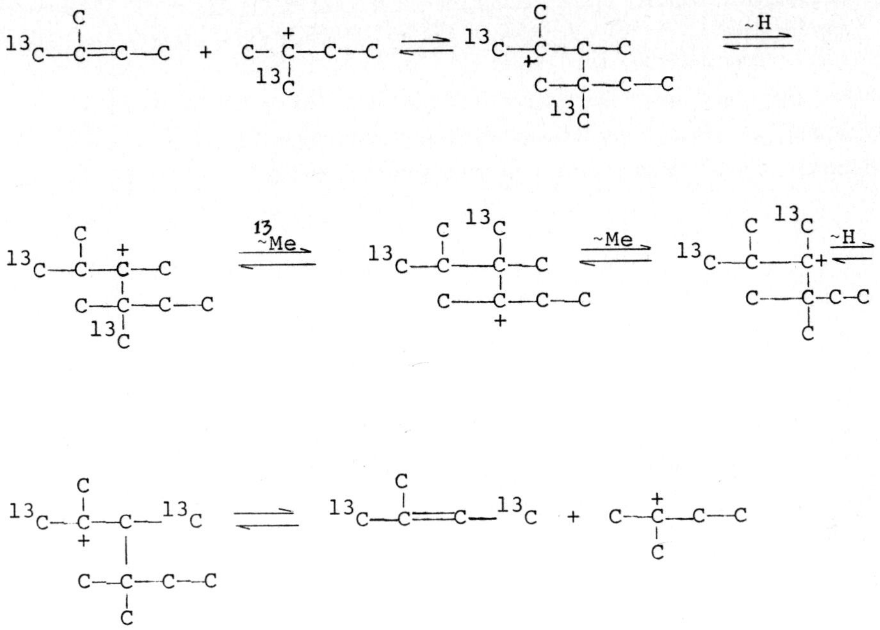

SCHEME 5

Additionally, the data indicated that the six methyl carbons of the C_{10} cation were scrambled, during the course of the reactions by rapid hydrogen and methyl shifts. The non-methyl carbons of the alkyl chain (C-2 and C-3 of tert-pentyl chloride) may not have attained statistical distribution.

Protonated cyclopropanes could not be intermediates in the isotopic rearrangement because such intermediates would lead to scrambling of the label between methyl and non-methyl carbon atoms (eq. 28) which was not observed.

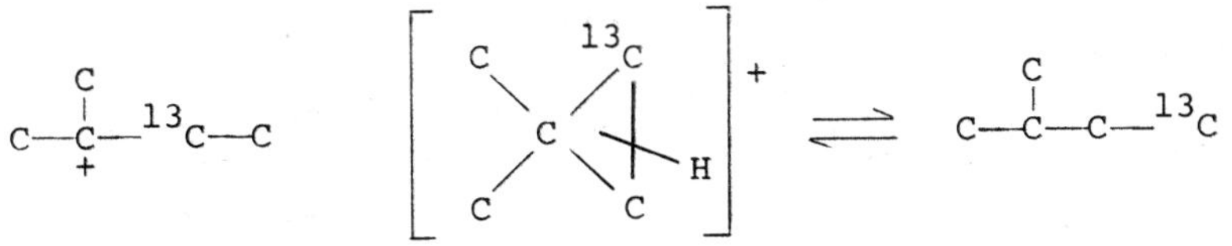

[28]

5. Rearrangements Accompanying a Cyclialkylation. Reaction of 3-chloro-4-fluoro-2-methylpropiophenone (5) with
AlCl$_3$ was reported to give three products: 5-fluoro-2-methyl-
indanone (6), 5-fluoro-3-methyl-1-indanone (7), and 2-(4'-
fluorophenyl)-1-oxoniacyclopent-1-enyl cation (8).[40a]
formation of the unexpected rearrangement products 7 and 8

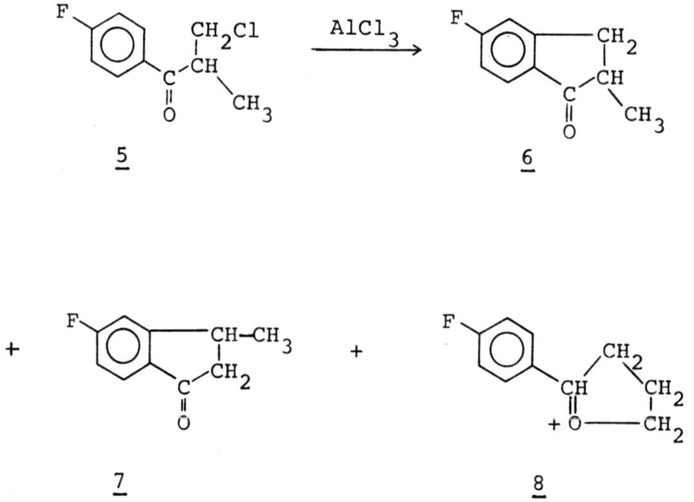

could be rationalized in terms of either alkyl (methyl) and
hydride shifts or an acyl migration. In order to distinguish
between these alternatives, Pines and Douglas prepared samples
of 5 which were labeled with ^{13}C at the carbon holding the
chlorine and with ^{2}H at the carbon α to the carbonyl group.[40b]
Experiments with the labeled molecules in which ^{2}H
and ^{13}C NMR analysis was employed excluded acyl migration as
a contributory pathway for the rearrangements. The investigators suggested an initial 1,2-methide shift preceding cyclialkylation to form 7 and hydride shifts following the initial
methide shift to account for the formation of 8.[40b]

6.   Isomerization and Condensation of Alkanes by Strong
Acids.   Since n-butane is not isomerized to isobutane by HF-
SbF$_5$,[41] whereas n-pentane and n-hexane are rapidly converted
under the same conditions into isopentane and isohexanes respectively,[42] a mechanism (eq. 30) involving protonated cyclopropane intermediates was proposed to explain the difference
in the behavior of butane and the higher alkanes.   According
to this mechanism, opening of the cyclic intermediate from n-
butane to produce isobutane would involve a primary carbocation, whereas isomerizations of n-pentane and n-hexane could

$$
\begin{array}{c}
\underset{+}{\overset{CH_3}{\underset{\scriptstyle}{}}}\!\!\diagdown\!\!\underset{\substack{CH_2 \\ \\ }}{\overset{}{}}\!\!\diagup \underset{\substack{\overset{|}{H}}}{CH\!-\!R} \;\rightleftharpoons\; CH_3\diagup\!\!\underset{\substack{\overset{|}{H}}}{C}\!\!\overset{CH_2}{\underset{H^+}{\diagup}}\!\!-\!CH\!-\!R \;\rightleftharpoons\; CH_3\!-\!\underset{+}{\overset{\overset{\displaystyle CH_3}{|}}{CH}}\!-\!CH\!-\!R
\end{array}
$$

$$
\rightleftharpoons\quad CH_3\!-\!\underset{+}{\overset{\overset{\displaystyle CH_3}{|}}{C}}\!-\!CH_2R \qquad R=H,\ CH_3,\ \text{or}\ C_2H_5 \qquad\qquad [30]
$$

proceed via secondary carbocations.   Although opening of the
protonated cyclopropane from n-butane to secondary carbocations could not produce isobutane, it could rearrange an isotopic lable from C-1 to C-2.   In 1968, Brouwer[42] treated a
mixture of n-butane-1-^{13}C and n-pentane with methylcyclopentyl
ions prepared from methylcyclopentane and HF-SbF$_5$.   The ratio
of n-butane-1-^{13}C to n-butane-2 ^{13}C was determined mass-
spectrometrically by comparing the peaks at 43 and 44 amu.   It
was found that n-butane-1-^{13}C was isomerized to n-butane-2-^{13}C
at a rate comparable to the isomerization of n-pentane to isopentane.   No significant amount of isobutane was formed.
These results were explained in terms of the mechanism involving a protonated cyclopropane intermediate which can open to a
secondary but not a primary carbocation (eq. 31).   However, it
should be noted that in the presence of aluminum bromide and a

$$
\begin{array}{c}
\text{CH}_3 \quad \text{CH}_2 \\
\diagdown \quad \diagup \\
\text{CH} \quad ^{13}\text{CH}_3 \\
\text{+}
\end{array}
\quad \longrightarrow \quad
\begin{array}{c}
\text{CH}_3 \quad \text{CH}_2 \\
\diagdown \quad | \\
\text{C} \quad\quad ^{13}\text{CH}_3 \\
\diagup \quad | \\
\text{H} \quad\quad \text{H}^+
\end{array}
\qquad [31]
$$

cocatalyst (H_2O), the isomerization of n-butane to isobutane and the scrambling of ^{13}C-label proceed at the same rate.[44] This isomerization of n-butane, as well as the isotopic isomerization of propane-1-^{13}C, was found to be completely intramolecular.[44, 45] Another mechanism involving different behavior of protonated cyclopropanes or bond to bond rearrangements, as suggested by Olah and coworkers,[46, 47] may account for the differences.

Olah and coworkers[48-50] made a number of highly interesting studies of the reactions of alkanes and alkylbenzenes in FSO_3H-SbF_5, HF-SbF_5 and related superacid media. For example, when isobutane was treated with DSO_3F-SbF_5 or DF-SbF_5 at atmospheric pressure and $-78°$, there occurred substantial exchange between the methine proton and deuterium as determined by mass spectrometry and ^{2}H-nmr and ^{1}H-nmr of the recovered isobutane, but the methyl protons underwent much less exchange. Some isomerization to n-butane and cleavage to CH_3D and propane also took place. Starting from either isobutane or n-butane, in the presence of HF-SbF_5, an equilibrium mixture of the two isomers was formed as well as other products. The isomerization was believed to involve a protonated methylcyclopropane intermediate.[15] These results were explained by the researchers on the basis of a mechanism involving frontside electrophilic attack by a proton on a C-H or C-C bond to produce a three-center penta-coordinated transition state (designated a carbonium ion by Olah and coworkers because it is in its highest coordination state). Subsequently the three center transition state can collapse to the original alkane and H^+, or form a trivalent sp^2 carbocation (designated by them as a carbenium ion) and H_2 or, if the attack is at the C-C bond, form a lower alkane and a carbocation by cleavage. A formulation of this proposed mechanism is shown in Scheme 6.

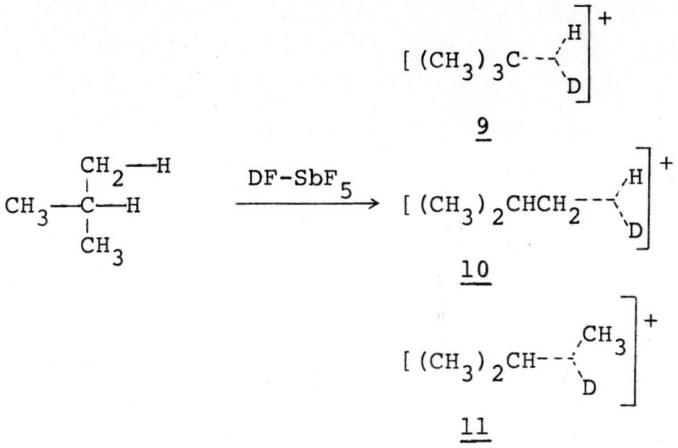

$$\left[(CH_3)_3C\text{-}\text{-}\overset{\overset{\displaystyle H}{\diagup}}{\underset{\diagdown}{}}D \right]^+$$

9

$$CH_3\text{-}\overset{\overset{\displaystyle CH_2\text{-}H}{|}}{\underset{\underset{\displaystyle CH_3}{|}}{C}}\text{-}H \quad \xrightarrow{\text{DF-SbF}_5} \quad \left[(CH_3)_2CHCH_2\text{-}\text{-}\overset{\overset{\displaystyle H}{\diagup}}{\underset{\diagdown}{}}D \right]^+ \qquad [32]$$

10

$$\left[(CH_3)_2CH\text{-}\text{-}\overset{\overset{\displaystyle CH_3}{\diagup}}{\underset{\diagdown}{}}D \right]^+$$

11

$$(\underline{9}) \;\rightleftharpoons\; (CH_3)_3C^+ + HD \qquad\qquad\qquad [33]$$

$$\rightleftharpoons\; (CH_3)_3CD + H^+$$

$$(\underline{10}) \;\rightleftharpoons\; (CH_3)_2CHCH_2^+ \;\xrightarrow{\;\sim Me\;}\; CH_3\overset{+}{C}HCH_2CH_3 \qquad [34]$$

$$\downarrow RH$$

$$\rightleftharpoons\; (CH_3)_2CHCH_2D \qquad\qquad CH_3CH_2CH_2CH_3$$

$$(\underline{11}) \;\longrightarrow\; (CH_3)_2CH^+ + CH_3D \qquad\qquad [35]$$

$$RH \downarrow$$

$$CH_3CH_2CH_3$$

Scheme 6

In contrast to the findings of Olah and coworkers, Otvos[51] found that in the reaction of isobutane with D_2SO_4 the nine methyl protons underwent deuterium exchange but not the methine hydrogen. This exchange presumably occurred through an iso-butylene intermediate.

Further support for the involvement of frontside attack and three-center transition states was obtained by the treatment of adamantane with DF-SbF$_5$ (eq. 36). Deuterium was found mainly at the bridgehead positions (by ^{2}H-nmr, ^{1}H-nmr, and mass spectrometry). In adamantane elimination to an olefin and backside attack are both ruled out by the rigid molecular structure.[15]

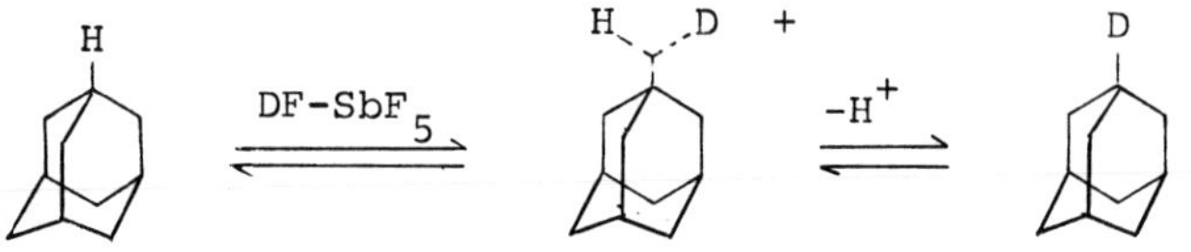

$$[36]$$

7. Alkylation and Polycondensation of Alkanes. The formation of higher molecular weight alkanes by electrophilic attack of a carbocation on a C-H or C-C single bond has been demonstrated by the polycondensation of methane to form ethane and by the condensation of methane, ethane, propane, or butanes to form highly branched polyalkanes of low molecular weight when an excess of alkane was treated with $FSO_3H\text{-}SbF_5$ ("magic acid") (eq. 37).[46, 49]

$C_4H_9^+$, etc. $$[37]$$

The yields of alkylation products were often quite low.* For example, the methylation of methane by the $CH_3F\text{-}SbF_5$ complex in the $30°\text{-}50°$ temperature range produced less than 3% of alkylate, mostly C_4, C_6, and some C_8. When $CD_3F\text{-}SbF_5$ was used, hydrogen transfer was observed forming CD_3H as the major product.

When isobutane was allowed to react with the tert-butyl cation, only a trace of 2,2,3,3-tetramethylbutane was formed, but this trace is good evidence of the ability of the tert-

---

*The propylation of propane by isopropyl cation gave C_6 alkylates in about 20% yield based on the starting cation, however.

butyl cation to attack a C-H single bond in spite of steric hindrance (eq. 38).[46, 49]

$$(CH_3)_3C\text{—}H + {}^+C(CH_3)_3 \rightleftharpoons (CH_3)_3C\text{—}C(CH_3)_3 + H^+ \qquad [38]$$

C. Alkylbenzene Rearrangements

1. _Ethylbenzene._ In 1955, Roberts, Ropp, and Neville reported that treatment of ethyl-β-^{14}C-benzene with aluminum chloride under a number of different conditions gave no rearrangement to ethyl-α-^{14}C-benzene.[7] Alkylation of benzene with ethyl-β-^{14}C chloride gave exclusively ethyl-β-^{14}C-benzene. However, treatment of ethyl-β-^{14}C chloride with aluminum chloride gave almost complete equilibration in one hour between ethyl-α-^{14}C chloride and ethyl-β-^{14}C chloride (cf. the previous section on rearrangement of ethyl halides). Later it was reported that three consecutive treatments with fresh aluminum chloride of ethyl-β-^{14}C-benzene at $100°$ for 6.5 hours each time produced 8% isotopic rearrangement.[52]

2. _n-Propylbenzene._ In 1957, Roberts and Brandenberger reported the first example of isotopic rearrangement in an alkylbenzene recovered after treatment with aluminum chloride. Heating labeled n-propylbenzene with aluminum chloride for 6.5 hr at $100°$ caused the α- and β-carbon atoms to interchange, the γ-carbon atom not being affected. Thus treatment of one mole of n-propyl-β-^{14}C-benzene with 0.33 mole aluminum chloride for 6.5 hr at $100°$ caused 31% of the carbon-14 activity to migrate to the α-position.[53]

By using more powerful complex acids as catalysts in multiple treatments, in which the recovered n-propylbenzene from one treatment was retreated with fresh catalyst, Roberts and Douglass were able to increase the percentage migration to 48% (percentage ^{14}C rearranged from C-α to C-β or vice versa).[54] Roberts and Greene reported the most efficient rearrangement conditions in which benzene was employed as solvent and a small amount of water as cocatalyst.[55, 56] Under these conditions, complete equilibration of ^{14}C between the α- and β-carbons was achieved in a single catalyst treatment.

In all of these reactions, the treatment of n-propylbenzene

with aluminum chloride gave only 2-5% isomerization to isopropylbenzene. Disproportionation of the alkylbenzenes to the
corresponsing dialkylbenzenes also occurred, and the dialkylbenzenes were found to have the same degree of isotopic rearrangement in their side chains.[53, 54] In order to explain the
observed phenomena, especially the isotopic rearrangement, a
variety of mechanisms were put forward. Roberts and
Brandenberger,[53] Nenitzescu,[57] and Roberts and Douglass[58]
proposed mechanisms in which a methide shift occurred to give
either a protonated complex, a primary carbocation, or a
phenonium ion equivalent to at least an incipient primary carbocation (Scheme 7). Such a pathway is energetically unfavorable. Moreover, an article by Karabatsos and Vane[39] had

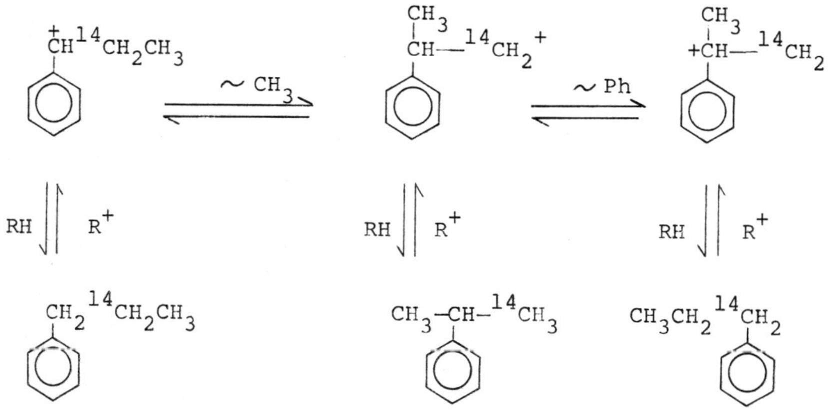

Scheme 7

brought to prominence the fact that at least one reaction
hitherto thought to involve primary carbocations in fact involved dimerization, rearrangement of the dimer, and scission
to give products, via mechanisms in which only secondary or
tertiary carbocations were involved.

These facts suggested the question: Could scrambling of
the α- and β-carbon atoms of n-propylbenzene occur in such a
dimer? In 1964 Farcasiu[59] proposed such an intermolecular or
"bimolecular" mechanism involving dimeric cationic species to

account for the movement of the label from the β- to the α-
position in n-propyl-β-^{14}C-benzene (Scheme 8). It may be seen
that none of the intermediate diphenylhexyl cations were primary.

Earlier in the same year that Farcasiu proposed his "bimolecular" mechanism, a new mechanism involving diarylpropanes
as key intermediates was proposed by Roberts, Khalaf, and
Greene[55, 56] and independently by Streitwieser and Reif[60] to
account for these side chain rearrangements. This mechanism
also avoids the involvement of primary carbocations. It re-
ceived experimental support from the isolation of diaryl-
alkanes from reactions of propyl- and butylbenzenes with $AlCl_3$

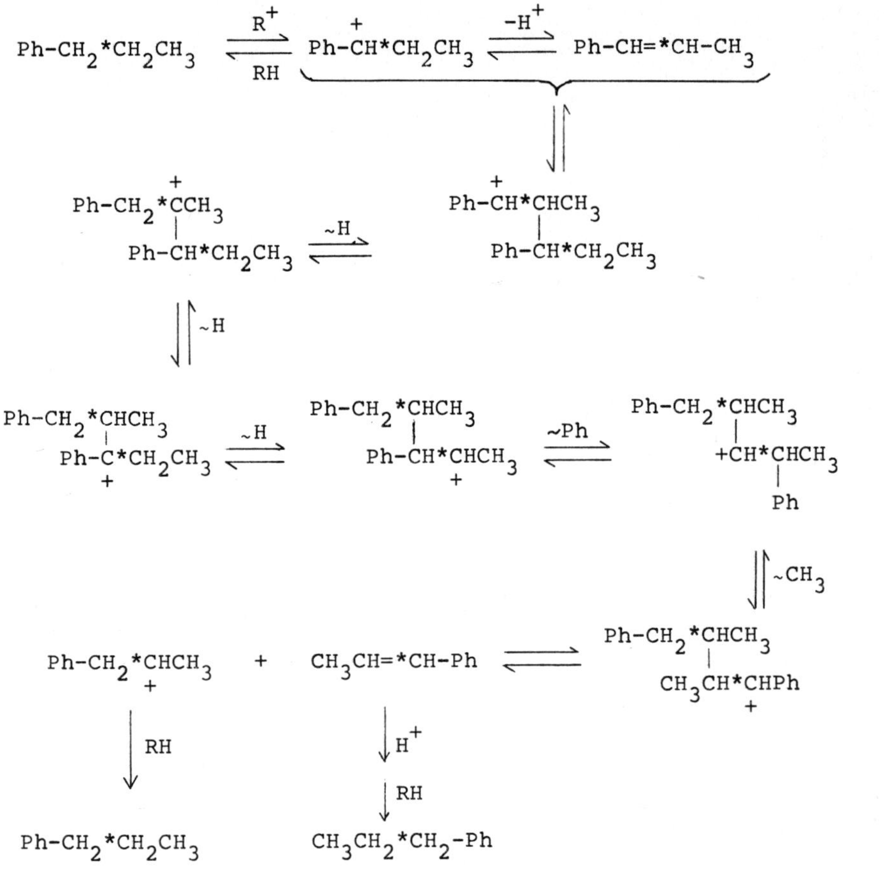

$$Ph-CH_2\text{*}CH_2CH_3 \underset{RH}{\overset{R^+}{\rightleftharpoons}} Ph-\overset{+}{C}H\text{*}CH_2CH_3 \overset{-H^+}{\rightleftharpoons} Ph-CH=\text{*}CH-CH_3$$

$$\begin{array}{c} Ph-CH_2\text{*}\overset{+}{C}CH_3 \\ | \\ Ph-CH\text{*}CH_2CH_3 \end{array} \overset{\sim H}{\rightleftharpoons} \begin{array}{c} Ph-\overset{+}{C}H\text{*}CHCH_3 \\ | \\ Ph-CH\text{*}CH_2CH_3 \end{array}$$

$$\Big\Updownarrow \sim H$$

$$\begin{array}{c} Ph-CH_2\text{*}CHCH_3 \\ | \\ Ph-\overset{+}{C}\text{*}CH_2CH_3 \end{array} \overset{\sim H}{\rightleftharpoons} \begin{array}{c} Ph-CH_2\text{*}CHCH_3 \\ | \\ Ph-CH\text{*}\overset{+}{C}HCH_3 \end{array} \overset{\sim Ph}{\rightleftharpoons} \begin{array}{c} Ph-CH_2\text{*}CHCH_3 \\ | \\ +CH\text{*}CHCH_3 \\ | \\ Ph \end{array}$$

$$\Big\Updownarrow \sim CH_3$$

$$\begin{array}{c} Ph-CH_2\text{*}\overset{+}{C}HCH_3 \\ \end{array} + CH_3CH=\text{*}CH-Ph \rightleftharpoons \begin{array}{c} Ph-CH_2\text{*}CHCH_3 \\ | \\ CH_3CH\text{*}\overset{+}{C}HPh \end{array}$$

$$\Big\downarrow RH \qquad \Big\downarrow H^+$$

$$\Big\downarrow RH$$

$$Ph-CH_2\text{*}CH_2CH_3 \qquad CH_3CH_2\text{*}CH_2-Ph$$

Scheme 8

by Roberts and co-workers,[55, 56] who also demonstrated that complete equilibration of ^{14}C between the α- and β-position in n-propyl-α-^{14}C-benzene was easily achieved under conditions that favor the formation of diphenylpropanes.[55] A formulation of this mechanism is given below (Scheme 9). Ar can represent either a phenyl or a substituted phenyl group because of the observed rapid alkylation-dealkylation reactions.[61, 62]

When the n-propyl-α-^{14}C-benzene was treated with aluminum chloride at 100° in the presence of increasing amounts of added benzene, it was found that the amount of isotopic rearrangement of the n-propylbenzene increased and then decreased, and that the concentrations of the diphenylpropanes also increased and decreased in direct relation to the increase and decrease of isotopic rearrangement.[56] In subsequent work, the proposed intermediates in the isotopic rearrangement of n-propyl-α-^{14}C-benzene, 1,1-diphenyl-propane-α-^{14}C, and 1,2-diphenylpropane-α-^{14}C were actually prepared and subjected to treatment with aluminum chloride.[54] It was found that the ^{14}C-label became equilibrated between the α- and β-positions of the treated and recovered 1,2-diphenylpropane-^{14}C even at room temperature. The labeled n-propyl-^{14}C-benzene formed by dealkylation from 1,2-diphenylpropane also showed equilibration of ^{14}C between

$$Ar^*CH_2CH_2CH_3 \underset{}{\overset{R^+}{\rightleftharpoons}} Ar^*CH_2\overset{+}{C}H\text{-}CH_3 \underset{}{\overset{ArH}{\rightleftharpoons}} Ar^*CH_2\underset{\underset{Ar}{|}}{C}HCH_3$$

$$\Bigg\updownarrow R^+$$

$$ArCH_2^*CH\text{-}CH_3 \underset{RH}{\overset{}{\rightleftharpoons}} Ar^*\overset{\overset{CH_3}{|}}{C}H\text{-}\overset{+}{C}H \underset{}{\overset{\sim CH_3}{\rightleftharpoons}} Ar^*\overset{+}{C}H\text{-}CHCH_3$$
$$\underset{Ar}{|} \qquad\qquad \underset{Ar}{|} \qquad\qquad \underset{Ar}{|}$$

$$\Bigg\updownarrow ArH \qquad \times \qquad [^+CH_2^*CHCH_3]$$
$$\qquad\qquad\qquad\qquad \underset{Ar}{|}$$

$$\underset{RH}{}$$
$$ArCH_2^*CH\text{-}CH_3 \overset{RH}{\underset{}{\rightleftharpoons}} ArCH_2^*CH_2CH_3$$
$$+$$

Scheme 9

the α- and β-carbons of the side chain. Isotopic rearrangement did not occur in 1,1-diphenylpropane-α-^{14}C at room temperature, but at 80° there was interconversion of 1,1-diphenylpropane-α-^{14}C and 1,2-diphenylpropane-^{14}C and equilibration of the label in both compounds. In none of the diphenylpropane experiments was any label found in the γ-position, in agreement with the results for n-propyl-^{14}C-benzene. Therefore, all the results from experiments with ^{14}C-labeled diphenylpropanes supported the theory that they function as intermediates in the isotopic rearrangement of n-propylbenzene.

Thus, two plausible mechanisms had been proposed for the isotopic rearrangement of labeled n-propylbenzene, and further evidence was needed to choose between them: the diphenylpropane mechanism and the bimolecular mechanism proposed by Farcasiu.

Both of these mechanisms provide explanations for an important aspect of the n-propylbenzene isotopic rearrangement: the lack of rearrangement of the isotopic carbon to the γ-position. In both of them the intermediate carbocations which lead to equilibration of the isotope between the α- and β-positions can be secondary, whereas primary carbocations are required to put the isotope into the γ-position.[*]

An obvious and elegant test to distinguish between these two mechanisms would be to use ^{13}C as the labeling isotope. The diphenylhexyl cation mechanism proposed by Farcasiu would allow the formation of dilabeled and unlabeled as well as monolabeled molecules of n-propylbenzene, whereas only monolabeled isotopic isomers could result from rearrangement via diphenylpropane intermediates.

This test was carried out by Roberts and Gibson, who performed a number of experiments involving treatment of 21-42% monolabeled n-propyl-α-^{13}C-benzene with aluminum chloride or

---

*One deficiency of the "bimolecular mechanism" is the fact that it predicts the rearrangement of n-propylbenzene to isopropylbenzene as easily (i.e., via secondary and tertiary carbocations) as the isotopic rearrangement, in disagreement with experimental facts. The mechanism involving diphenylpropane intermediates allows for rearrangement to isopropylbenzene only via primary carbocations which would not be expected to be facile, in accord with experimental facts.

aluminum bromide as catalyst, water as cocatalyst, and benzene as solvent, at reflux for 7 to 16 hours.[63, 64] Before and after each of these treatments, the recovered ^{13}C-labeled propylbenzene samples were analyzed by mass spectrometry and in some cases by ^{13}C-nmr and proton nmr. Comparison of the parent peaks and tropylium ion peaks of labeled and unlabeled molecules in the samples indicated that the ratio of monolabeled molecules to unlabeled molecules did not change and no dilabeled molecules were observed. In the course of the reaction, 50% of the ^{13}C-label rearranged from the α-position to the β-position in the propylbenzene chain.

These data effectively ruled out Farcasiu's "bimolecular" mechanism for the isotopic rearrangement in n-propylbenzene.* On the other hand, the observed results would be expected according to the diphenylpropane mechanism.

3. n-Butylbenzene. Treatment with aluminum chloride at 100° for 3 hr with no solvent gave only 4-5% isotopic rearrangement of n-butyl-α-^{14}C-benzene, i.e., 4-5% loss of ^{14}C from the β-position.[52] Disproportionation was observed but no detectable isomerization to isobutylbenzene and sec-butylbenzene. More extensive tests of the possible isotopic rearrangement of n-butylbenzene were later made employing ^{13}C-labeling. These experiments confirmed that treatment of n-butyl-α-^{13}C-benzene with aluminum chloride or bromide in benzene under the same conditions used in the rearrangement of labeled n-propylbenzene produced no isotopic rearrangement within experimental error in the recovered n-butylbenzene. Under drastic conditions most of the n-butylbenzene was converted to 1-methylindane, di- and tri-n-butylbenzenes, small amounts of sec-butylbenzene, isobutylbenzene, and ethylbenzene.[63]

In order better to understand the reasons for the differ-

---

*It should be noted that although the ^{13}C-labeling experiments rule out diphenylhexyl cations as playing an important role in the equilibration of the α- and β-carbons in n-propylbenzene, it does not exclude the possibility that such dimeric cations may be the intermediates which lead to the small amounts of substituted indans and tetralins observed as by-products in the reactions of alkylbenzenes with $AlCl_3$.

ence in reactivity of n-butylbenzene and n-propylbenzene, 1,2-diphenylbutane-2-^{13}C was treated with anhydrous aluminum chloride in benzene at reflux for either 1 or 7 hours. The position of ^{13}C-label in the recovered 1,2-diphenylbutane and n-butylbenzene product was determined by ^{13}C-nmr and mass spectrometry, respectively.[65, 66] Isotopic rearrangement of 8-10% of the label from C-2 to C-1 was observed in the n-butylbenzene and 8% in the recovered 1,2-diphenylbutane. Therefore, under conditions which produced complete isotopic rearrangement between C-1 and C-2 in labeled 1,2-diphenyl-propane,[55] 1,2-diphenylbutane-2-^{13}C showed only 8-10% rearrangement. These results indicate that the ethyl shift required for the isotopic rearrangement is very slow compared to the methyl shift in the rearrangement of labeled 1,2-diphenyl-propane (eq. 39). Apparently n-butylbenzene preferentially

$$C_6H_5-CH_2-^{13}\underset{|}{C}H-CH_2CH_3 \;\; \underset{\xrightarrow{+R^+ \atop -H^-}}{\rightleftharpoons} \;\; C_6H_5-\overset{+}{C}H-^{13}\underset{|}{C}H-CH_2CH_3$$

$$\underset{\text{shift}}{\overset{\sim CH_2CH_3}{\rightleftharpoons}} \;\; C_6H_5-\underset{|}{C}H-^{13}\overset{+}{C}H-C_6H_5 \;\; \underset{\xleftarrow{RH}}{\rightleftharpoons} \;\; C_6H_5-^{13}CH_2-\underset{|}{C}H-CH_2CH_3 \;[39]$$

(with C_6H_5 substituents shown on the second carbons and CH_2CH_3 on the third)

forms 1,3-diphenylbutane which cyclizes and subsequently produces 1-methylindane rather than following the above pathway.

4. Reactions in "Super-Acid" Systems. Olah and co-workers[47] have observed by nmr spectroscopy the formation of a number of stable alkylbenzenium ions (12a-d) in HF-SbF$_5$, generally using an SO$_2$ClF or SO$_2$ClF-SO$_2$F solvent system at low temperatures (eq. 40). The 4-alkylbenzenium ion (12a) is the

12a   12b   12c   12d   [40]

most stable isomer and is "frozen out" at low temperatures, whereas a mixture of all the isomeric ions is observed at room temperature.

Reaction of toluene with excess DF-SbF$_5$ in SO$_2$ClF at -30°
for 3 hours incorporated 1.1 atom of deuterium per molecule
with statistical distribution of deuterium around the ring
carbons. These results indicated that intermolecular exchange
of deuterium between toluene and DF-SbF$_5$ is slow (eq. 41), but
the intramolecular exchange process is relatively fast (eq.
42).

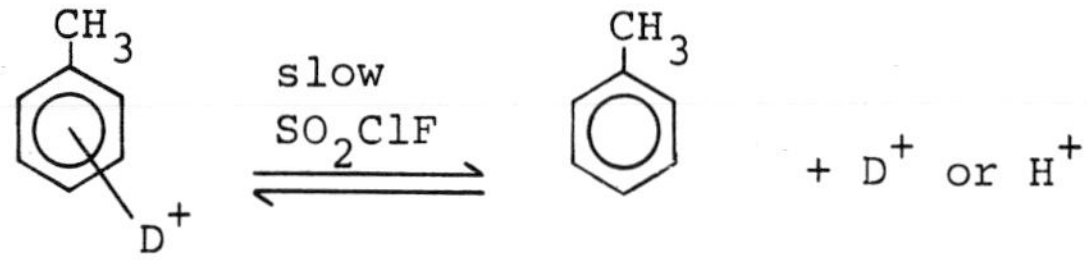

$$\text{[41]}$$

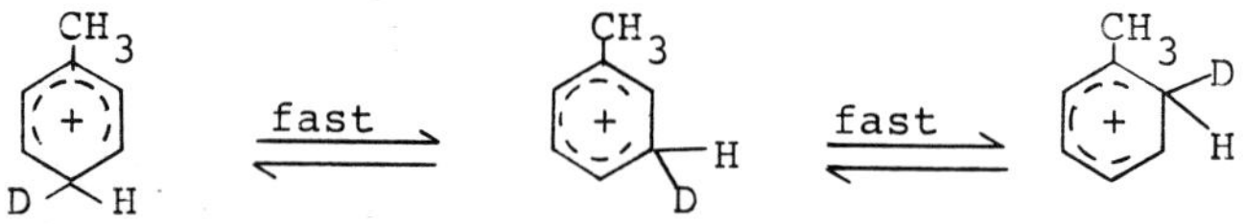

$$\text{[42]}$$

The actual details of the nmr assignments are beyond the
scope of this review, but certain of the conclusions presented
regarding reactions of alkylbenzenes with Lewis acid catalysts
may be noted.

Treatment of n-propylbenzene with HF-SbF$_5$ (1.1 molar) in
SO$_2$ClF (1-2 vol.) at -78° produced the n-propylbenzenium ions
only. When the mixture was warmed to +25° and kept there for
1.5 hr, 80% of the n-propylbenzenium ion remained but a small
amount of the phenyldimethylcarbonium ion was formed (eq. 43).
Barely detectable cleavage to produce benzenium ion was also

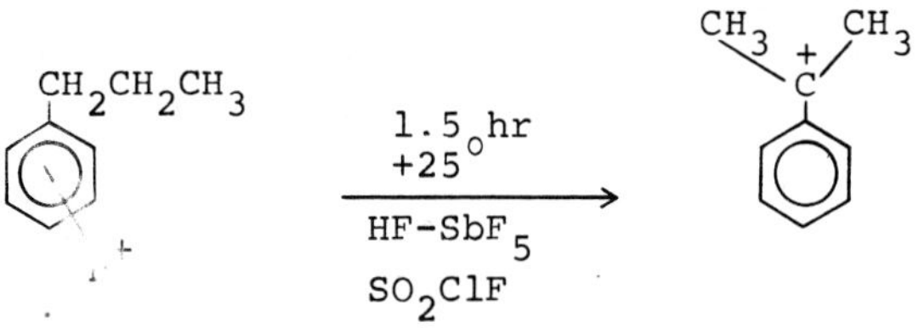

$$\text{[43]}$$

observed under these conditions. The formation of only a
small amount of phenyldimethylcarbenium ion corresponded
closely to the result when n-propylbenzene is treated with
aluminum chloride and heated, which produces at most 5% isomerization to isopropylbenzene.

When butylbenzenes were treated with HF-SbF$_5$ or HSO$_3$F-SbF$_5$
under the same conditions as the propylbenzenes, the butyl-
benzenium ions were easily prepared and were stable at low

temperatures. At room temperature the n-butylbenzenium ion remains stable (eq. 44), but the isobutylbenzenium ion undergoes about 15% rearrangement to the phenylethylmethylcarbenium ion (eq. 45), the sec-butylbenzenium ion undergoes up to 80% rearrangement to the isobutylbenzenium ion (eq. 46), and the tert-butylbenzenium ion dealkylates to form benzenium and tert-butyl ions (eq. 47).

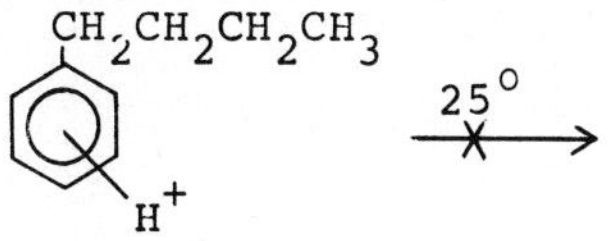

[44]

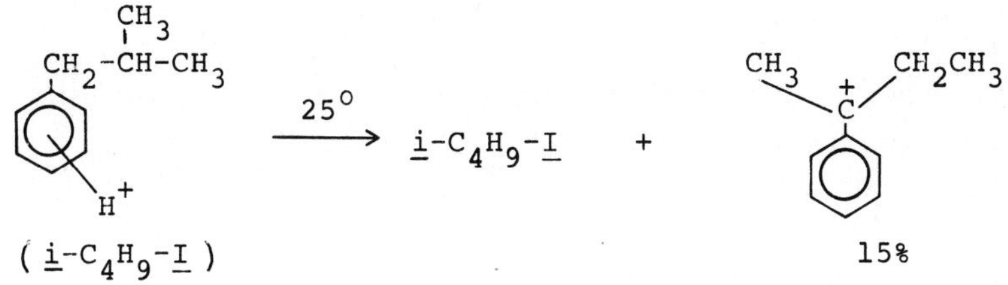

[45]

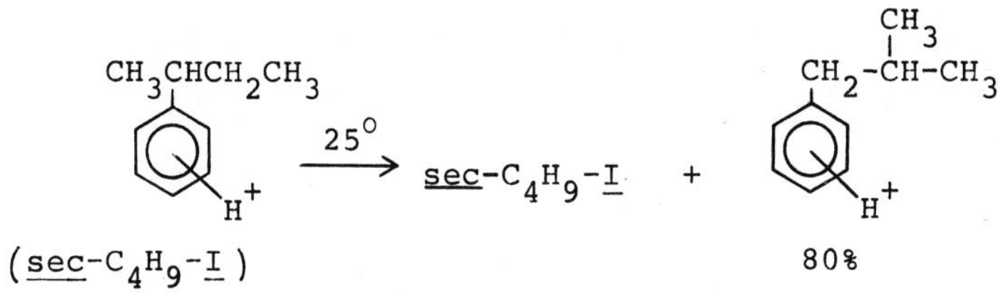

15%

[46]

80%

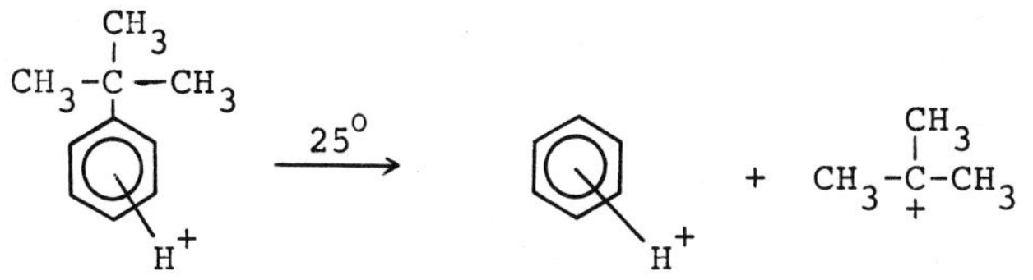

[47]

Olah and coworkers have also advanced a new mechanism to explain the rearrangement and dealkylation reactions of alkyl-benzenes.[47] This mechanism involves the protonation of the aromatic ring to form alkylbenzenium ions which can then iso-merize to three-center bonded species in which C-C single

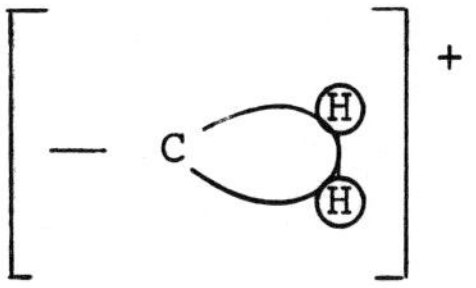

or

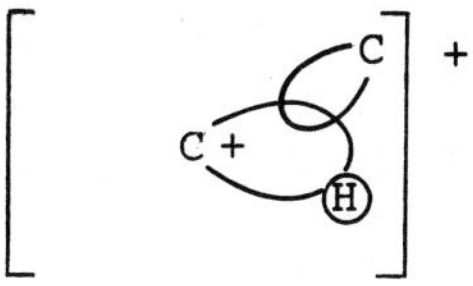

13   14

bonds of the side chain are involved; i.e., bonds involving overlap of one sp^3 orbital of carbon with two other bonding orbitals. Bond to bond rearrangement interconverting different three center-bonded species followed by deprotonation can then occur producing overall isomerization of the starting alkylbenzene (Scheme 10).

It was specifically claimed that the three-center bond to bond rearrangement can account for the isotopic rearrangement of n-propyl-α-^{14}C-benzene and the isomerization of sec-butyl-benzene to isobutylbenzene. The validity of the new mechanism in the case of the isotopic rearrangement of labeled n-propyl-benzene is, however, very questionable since it would clearly allow scrambling of label between the α, β, and γ carbons whereas only the α and β carbons are observed to be interchanged. It might however, be useful in explaining the formation of the small percentage of isopropyl benzene which is in fact observed.

Similarly, the mechanism of Olah and coworkers would lead one to expect not only the interconversion of sec-butylbenzene and isobutylbenzene under strong acid conditions but also interconversion of all the isomeric butylbenzenes which does not proceed beyond a very slight extent.[57, 67]

It should be noted that the mechanism involving three-center bonds and bond to bond rearrangements has the same general effect as the original mechanism proposed by Roberts and Brandenberger,[52] which suggested that protonation of labeled n-propylbenzene could form a π-complex in which a methide shift could occur to rearrange the label, and is also similar to protonated cyclopropane mechanisms. All of these mechanisms have the effect of allowing a rearrangement which would otherwise involve a primary carbocation. Even the incipient primary carbocations must require more energy for their formation than the pathway involving diarylalkane intermediates, which has the advantage that such intermediates have been observed under the reaction conditions. Thus the diaryl-alkane mechanism proposed by Roberts and coworkers[56] and by Streitweiser and Reif[60] remains the most likely mechanism for the isotopic rearrangement of labeled n-propylbenzene. This

Scheme 10

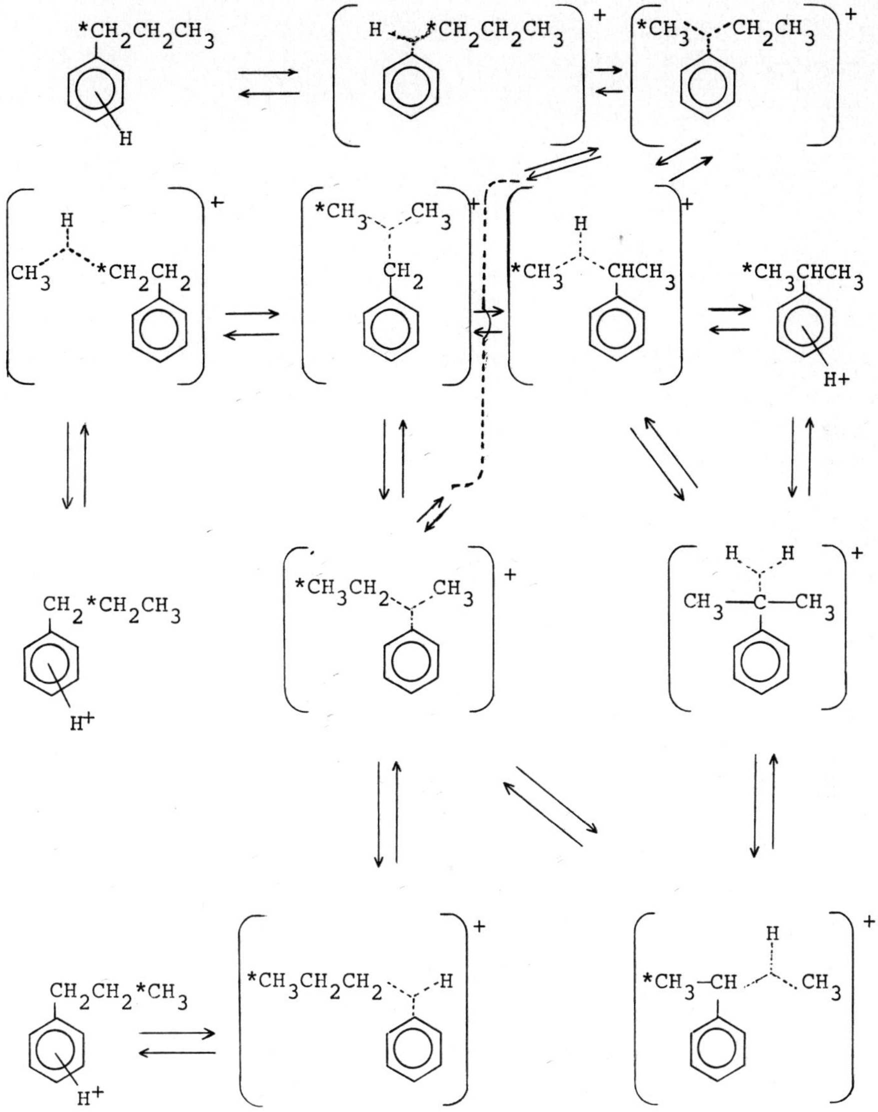
*CH2CH2CH3
H
H
*CH2CH2CH3
*CH3
CH2CH3
H
CH3
*CH2CH2
*CH3
CH3
CH2
H
*CH3
CHCH3
*CH3CHCH3
H+
CH2*CH2CH3
H+
*CH3CH2
CH3
H
H
CH3
C
CH3
CH2CH2*CH3
H+
*CH3CH2CH2
H
H
*CH3
CH
CH3

mechanism may also provide a pathway for the interconversion of _sec_-butylbenzene and isobutylbenzene[67] (Scheme 11), but this rearrangement may also proceed via the simpler pathway of Scheme 12 in which no primary carbonium ions appear.

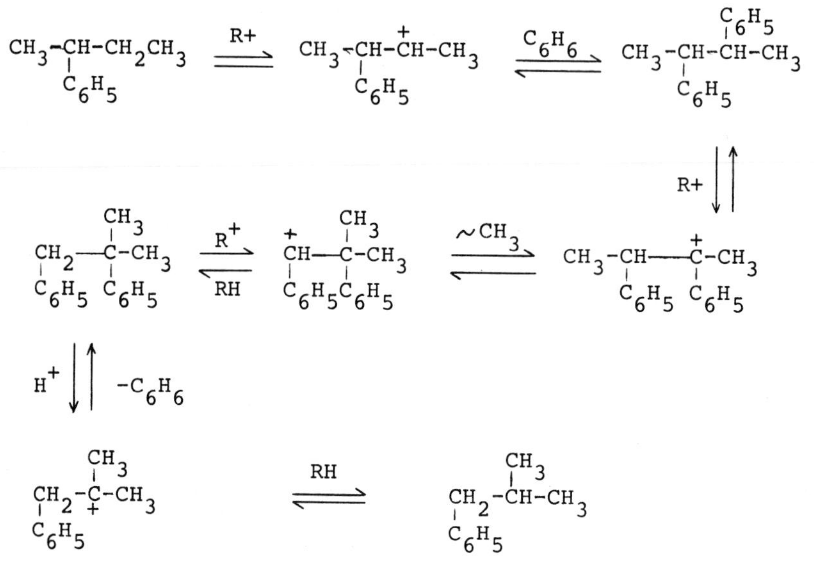

$$CH_3-CH-CH_2CH_3 \xrightarrow{R+} CH_3-CH-\overset{+}{C}H-CH_3 \underset{}{\overset{C_6H_6}{\rightleftarrows}} CH_3-CH-CH-CH_3$$

with C_6H_5 substituents, and C_6H_5 on the right product (with additional C_6H_5).

$$R+ \updownarrow$$

$$\underset{C_6H_5 \; C_6H_5}{\overset{CH_3}{CH_2-C-CH_3}} \underset{RH}{\overset{R^+}{\rightleftarrows}} \underset{C_6H_5C_6H_5}{\overset{CH_3}{\overset{+}{C}H-C-CH_3}} \overset{\sim CH_3}{\rightleftarrows} \underset{C_6H_5 \; C_6H_5}{CH_3-CH-\overset{+}{C}-CH_3}$$

$$H^+ \updownarrow \; -C_6H_6$$

$$\underset{C_6H_5}{\overset{CH_3}{CH_2-\overset{+}{C}-CH_3}} \underset{}{\overset{RH}{\rightleftarrows}} \underset{C_6H_5}{\overset{CH_3}{CH_2-CH-CH_3}}$$

Scheme 11

$$CH_3-CH-CH_2CH_3 \underset{RH}{\overset{R^+}{\rightleftarrows}} CH_3-CH-\overset{+}{C}H-CH_3 \overset{\sim CH_3}{\rightleftarrows} \underset{C_6H_5}{\overset{CH_3}{\overset{+}{C}H-CH-CH_3}}$$

with C_6H_5 substituents

$$RH \updownarrow R^+$$

$$\underset{C_6H_5}{\overset{CH_3}{CH_2-CH-CH_3}}$$

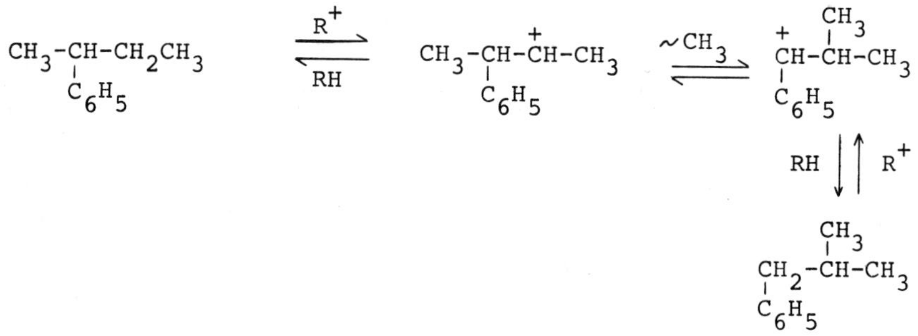

Scheme 12

One other experiment carried out with the aim of demonstrating rearrangement of an ethyl halide before alkylating benzene may be mentioned here because of its relation to the alkylbenzene rearrangement. Lee and coworkers[68] proposed that in an inert solvent such as 1,2,4-trichlorobenzene the alkyla-

tion rate might be decreased sufficiently to allow the $AlCl_3$-
induced rearrangement of ethyl-2-^{14}C iodide to take place
before it alkylated benzene. However, it was found that
significant amounts of ethyl-1-^{14}C-benzene were observed only
in reactions effected at high temperatures for extended times.
It was then demonstrated that the isotopic rearrangement
actually occurred subsequent to the initial formation of
ethyl-2-^{14}C-benzene. The isolation of the fragmentation
product toluene, having a molecular radioactivity 1/2 of that
of the original ethyl-2-^{14}C iodide, was interpreted as evidence for the formation of 1,2-diphenylethane, which func-
tioned as an intermediate in the isotopic rearrangement of the
ethylbenzene as well as for the fragmentation reaction that
produced toluene.

This suggestion is in good agreement with an unpublished
observation of Greene[55a] in connection with studies on the
mechanism of the isotopic propylbenzene rearrangement   By
isotope-dilution technique it was possible to determine that
the ethylbenzene isolated as a fragmentation product from
1,2-diphenylpropane-2-^{14}C had a molecular radioactivity of
one-half of that of the original diphenylpropane   This would
be the result expected from cleavage of the diphenylpropane
after equilibration of the isotope between the two non-methyl
carbons (Scheme 13).

5. sec-Butylbenzene and Higher sec-Alkylbenzenes. Some
interesting rearrangements of alkylbenzenes by Friedel-Crafts
catalysts have included reactions of other sec-alkylbenzenes:
2- and 3-phenylpentanes are interconverted by aluminum
chloride at $25°$[69, 70, 71] and 2-phenyldodecane is isomerized
to an equilibrium mixture of 2-, 3-, 4-, 5-, and 6-phenyldode-
canes at $50°$.[72]

From these results, it might be expected that an analogous
rearrangement involving a phenyl shift between secondary
carbon atoms would occur in sec-butylbenzene, which could only
be detected in the case of an isotopically labeled molecule
such as 2-phenyl-2-^{14}C-butane. In fact, Roberts, Anderson,
and Doss[73] determined that with water activated aluminum
chloride at $25°$ rearrangement to 2-phenyl-3-^{14}C-butane

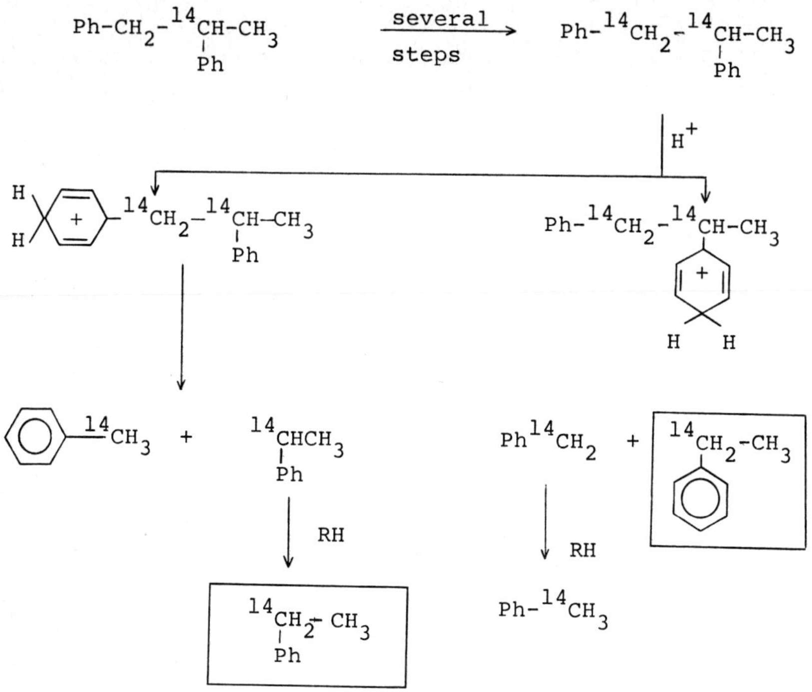

$$\text{Ph-CH}_2\text{-}^{14}\text{CH-CH}_3 \quad \xrightarrow[\text{steps}]{\text{several}} \quad \text{Ph-}^{14}\text{CH}_2\text{-}^{14}\text{CH-CH}_3$$
$$\underset{\text{Ph}}{|} \qquad\qquad\qquad\qquad\qquad \underset{\text{Ph}}{|}$$

Scheme 13

occurred, producing a 1:1 mixture of the two isotopic isomers in 1 hr. However, at -14 to -17°, with the same catalyst or with anhydrous aluminum bromide, no rearrangement occurred in 15 minutes; after 1 hr. at 0° with aluminum chloride-water, 21% rearrangement occurred. It is interesting also to compare the ease of the phenyl shift in sec-butylbenzene with its other reactions; for example, disproportionation to benzene and di-sec-butylbenzene (mainly meta and para isomers) was observed in all treatments of sec-butylbenzene with aluminum chloride or bromide from -17° to +25°. The isomerization of sec-butylbenzene to isobutylbenzene and vice versa giving a 1:2 ratio of the isomers happens only at higher temperatures (80 to 100°) as first reported in 1959.[57, 67]

D. Disproportionations and Reorientations of Alkylbenzenes

The alkyl groups in alkylbenzenes may migrate either intra-

molecularly or intermolecularly under the influence of
Friedel-Crafts catalysts. Intermolecular migration (trans-
alkylation) results in disproportionation; for example, ethylbenzene reacts rapidly with $AlCl_3$ at room temperature to produce a mixture of benzene, diethylbenzene, and higher-ethyl-
ated benzenes. o-Xylene is isomerized by $AlCl_3$ to a mixture
of all three xylenes. This isomerization may be called reorientation. Since the reorientation may occur with little or
no disproportionation to toluene and trimethylbenzenes, it
must take place by intramolecular migration of the methyl
groups. However, the reorientation of all alkyl groups larger
than methyl is accompanied by significant disproportionation,
so that the reorientation may involve both intra- and intermolecular alkyl transfers. Isotopes have been used in
attempts better to delineate the mechanisms of these dis-
proportionations and reorientations of alkylbenzenes.

Steinberg and Sixma[74] synthesized toluene-1-^{14}C and studied
its isotopic reorientation by $AlBr_3$ + HBr. They found that
the activity at carbon-1 diminished progressively towards 1/6
of the original value. The activity at carbon-2 (ortho) went
through a maximum and thereafter approached asymptotically the
value for statistical distribution, whereas the activities at
the meta and para positions showed a steady increase towards
the equilibrium values. A completely isomerized reaction mixture from a long reaction was shown by spectroscopic analysis
to contain only 7% m-xylene and 6% benzene, indicating that
intermolecular disproportionation was so much slower that it
did not interfere with the experimental analysis of the reorientation reaction.

Roberts and coworkers studied the question of possible
rearrangements in the alkyl side chain during disproportionation of ethyl-β-^{14}C-benzene and n-butyl-α-^{14}C-benzene in the
presence of aluminum chloride at 100.[7, 52] Degradation of the
side chains of both recovered alkylbenzenes and of both dialkylbenzenes indicated no significant rearrangement of the
[14]C-label. Thus labeled ethyl and butyl groups were shown to
undergo intermolecular transfer without rearrangement.

An elegant and more extensive study of the disproportionation of ethylbenzene was subsequently made by Streitwieser and Reif.[60] Optically active ethylbenzene-α-d-ring-^{14}C was prepared and subjected to disproportionation in a dilute benzene solution containing gallium bromide and hydrogen bromide. Recovered ethylbenzene was examined for radioactivity, optical rotation, and deuterium composition. Loss of ^{14}C activity and racemization occurred at equal rates, while deuterium scrambling was slightly slower. These results were interpreted in terms of a short-lived 1,1-diphenylethane intermediate formed by rapid and reversible addition of α-phenylethyl cation to benzene, the rate determining step being hydride exchange between the α-phenylethyl cation and the α-carbon of ethylbenzene. A mechanism of this type was considered likely to be general for disproportionation of primary alkylbenzenes, including, of course, n-butylbenzene.

Lack of rearrangement in the side chain of ethylbenzene was confirmed by Unseren and Wolf[75] using a doubly labeled molecule, ethyl-β-^{14}C-benzene-1-[14]C. Their main interest was in the reorientation of ethylbenzene-1-^{14}C by aluminum bromide-hydrogen bromide at 0°. They found the reorientated activity predominantly in the meta- and para-positions, which indicated that a rapid equilibration of the hydrocarbon and its localized pi-complex did not precede the reorientation reaction(s). Their results suggested that the reorientation proceeds predominantly via intermolecular alkyl transfer.

In 1965, Moore and Wolf made more extensive studies of the disproportionation and reorientation of ethylbenzene-1-^{14}C and performed a more exact determination of label position by stepwise degradation of the recovered ethylbenzene-[14]C.[76] The distribution of rearranged ^{14}C-label, mostly in the para-position with smaller amounts in the meta- and ortho-positions after short reaction times, suggested predominantly trans-alkylation followed by intramolecular shifts but did not exclude other possibilities. Further evidence supporting transalkylation as the major pathway of reorientation of ethyl groups has been obtained recently with [13]C- and [2]H-labeled ethylbenzene molecules.[77]

REFERENCES

1. F. Fairbrother, J. Chem. Soc., (1937) 503; ibid., (1941) 293.

2. G. Oulevey and B. P. Susz, Helv. Chim. Acta., 44, (1961) 1425.

3. E. L. Makor, P. J. Smit, and J H. Van der Walls, Trans. Faraday Soc., 53, (1957) 1309; A. I. Shatenshtein, Dokl. Akad. Nauk, SSSR, 91, (1953) 577.

4. (a) F. L. J. Sixma, and H. Hendriks, Koninkl. Ned. Akad. Wetenschap. Proc., Ser. B, 59, (1956) 61; (b) Rec. Trav. Chim., 75, (1956) 169.

5. F. L. J. Sixma, H. Hendriks, and D. Holtzapffel, Rec. Trav. Chim., 75, (1956) 127.

6. C. H. Wallace, and J. E. Willard, J. Am. Chem. Soc., 72, (1950) 5275.

7. R. M. Roberts, G. A. Ropp, and O. K. Neville, ibid., 77, (1955) 1764.

8. C. C. Lee, M. C. Hamblin, and N. James, Can. J. Chem., 36, (1958) 1597.

9. R. Nakane, O. Kurihara, and A. Natsubori, J. Am. Chem. Soc., 91, (1969) 4528.

10. A. Natsubori and R. Nakane, J. Org. Chem., 35, (1970) 3372.

11. B. J. Carter, W. D. Corey and F. D. DeHaan, J. Am. Chem. Soc., 97, (1975) 4783.

12. L. M. Stock and H. C. Brown, "Advances in Physical Organic Chemistry," Vol. I, V. Gold, Ed., Academic Press, Inc., New York, NY, 1963, pp. 35-154.

13. C. C. Lee, A. G. Foreman, and A. Rosenthal, Can. J. Chem., 35, (1957) 220.

14. M. A. McMahon and S. C. Bunce, J. Org. Chem., 29, (1964) 1515.

15. G. A. Olah, J. R. De Member, R. H. Schlosberg, and Y. Halpern, J. Am. Chem. Soc., 94, (1972) 156.

16. L. M. Nash, T. I. Taylor, and W. von E. Doering, J. Am. Chem. Soc., 71, (1949) 1516.

17. H. S. A. Douwes, and E. C. Kooyman, Rec. Trav. Chem. Pays-Bas, 83, (1964) 276.

142

18. G. J. Karabatsos, J. L. Fry, and S. Meyerson, Tetrahedron Lett., 38, (1967) 3735.
19. G. J. Karabatsos, J. L. Fry, and S. Meyerson, J. Am. Chem. Soc., 72, (1970) 614.
20. C. C. Lee and D. J. Woodcock, Can. J. Chem., 48, (1970) 858.
21. C. C. Lee and D. J. Woodcock, J. Am. Chem. Soc., 92, (1970) 5992.
22. J. D. Roberts and M. Halman, ibid., 75, (1953) 5759.
23. C. C. Lee, J. E. Kruger, and E. W. Wong, J. Am. Chem. Soc., 87, (1965) 3985.
24. I. L. Reich, A. Diaz, and S. Winstein, J. Am. Chem. Soc., 91, (1969) 5635.
25. C. C. Lee and J. E. Kruger, J. Org. Chem., 44, (1966) 2343; Tetrahedron 23, (1967) 2539.
26. G. J. Karabatsos, C. E. Orzech, and S. Meyerson, J. Am. Chem. Soc., 92, (1970) 606.
27. G. J. Karabatsos, C. E. Orzech, and S. Meyerson, J. Am. Chem. Soc., 87, (1965) 4394.
28. O. A. Reutov and T. N. Shatkina, Tetrahedron, 18, (1962) 237.
29. O. A. Reutov and T. N. Shatkina, Dokl. Akad. Nauk , SSSR, 133, (1960) 606.
30. V. Mayer and F. Forster, Chem. Ber., 9, (1876) 535.
31. D. M. Brouwer and H. Hogeveen in "Progress in Physical Organic Chemistry," A. Streitweiser, Jr. and R. W. Taft, Ed., Vol. 9, Wiley-Interscience Publishers, New York, NY, 1972, p. 179.
32. O. A. Reutov and T. N. Shatkina, Izv. Akad. Nauk., Ozd Khim. Nauk., (1963) 195.
33. O. A. Reutov, Pure Appl. Chem., 7, (1963) 203.
34. A. B. Susan, Studii Cercetari Chim., 14 (1966) 23.
35. C. C. Lee, B. Hahn, K. Wan, and D. J. Woodcock, J. Org. Chem., 34, (1969) 3210.
36. G. J. Karabatsos, C. Ziuodrou, and S. Meyerson, J. Am. Chem. Soc., 92, (1970) 5996.
37. J. L. Fry and G. J. Karabatsos in "Carbonium Ions," Vol. 2, G. A. Olah and P. von R. Schleyer, Ed., Interscience

New York, NY, 1970, Chap. 14.

38. J. D. Roberts, R. E. McMahon and J. S. Hine, J. Am. Chem. Soc., 73, (1950) 4237.

39. (a) G. J. Karabatsos, F. M. Vane, and S. Meyerson, ibid., 83, (1961) 4297; (b) G. J. Karabatsos and F. M. Vane, ibid., 85, (1963) 729; (c) G. J. Karabatsos, F. M. Vane, and S. Meyerson, ibid., 85, (1963) 733.

40. (a) S. H. Pines and A. W. Douglas, J. Am. Chem. Soc., 98, (1976) 8119; (b) J. Org. Chem., 43, (1978) 3126.

41. D. A. McCaulay, and A. P. Lien, Tetrahedron, 5, (1959) 186.

42. (a) D. M. Brouwer and J. M. Oelderik, Rec. Trav. Chim., 87, (1968) 721; (b) Preprints of papers presented before the Division of Petroleum Chemistry, ACS Meeting, San Francisco, April, 1968, p. 184.

43. D. M. Brouwer, Rec. Trav. Chim., 87, (1968) 1435.

44. J. W. Otvos, D. P. Stevenson, C. D. Wagner, and O. Beeck, J. Chem. Phys., 16, (1948) 745

45. O. Beeck, J. W. Otvos, D. P. Stevenson, and C. D. Wagner, J. Chem. Phys., 16, (1948) 255.

46. G. A. Olah and J. A. Olah, J. Am. Chem. Soc., 93, (1971) 1251.

47. G. A. Olah, R. H. Schlosberg, R. D. Porter, Y. K. Mo, D. P. Kelly, and G. D. Mateescu, ibid., 94, (1972) 2034.

48. G. A. Olah, J. R. De Member, and J. Shen, ibid., 95, (1973) 4952.

49. G. A. Olah, Y. Halpern, J. Shen, and Y. K. Mo, ibid., 95, (1973) 4960.

50. G. A. Olah, Y. K. Mo, and J. A. Olah, ibid., 95, (1973) 4934.

51. J. W. Otvos, D. P. Stevenson, C. D. Wagner, and O. Beeck, J. Am. Chem. Soc., 73, (1951) 5741; D. P. Stevenson, C. D. Wagner, O. Beeck, and J. W. Otvos, ibid., 74, (1952) 3269.

52. R. M. Roberts, S. G. Brandenberger, and S. G. Panayides, ibid., 80, (1958) 2507; R. M. Roberts, E. K. Baylis, and G. J. Fonken, ibid., 85, (1963) 3454.

53. R. M. Roberts, and S. G. Brandenberger, J. Am. Chem. Soc.,

144

$\underset{\sim}{79}$, (1957) 5484.

54. (a) R. M. Roberts and J. E. Douglass, Chem. and Ind.,
    (1958) 1557; (b) R. M. Roberts and J. E. Douglass, J.
    Org. Chem., $\underset{\sim}{28}$, (1963) 1225.

55. (a) R. N. Greene, Ph.D. Dissertation, University of Texas,
    1964; (b) R. M. Roberts and R. N. Greene, Acta Cient.
    Venezolana, $\underset{\sim}{15}$ [6] (1965) 251.

56. R. M. Roberts, A. A. Khalaf, and R. N. Greene, J. Am.
    Chem. Soc., $\underset{\sim}{86}$, (1964) 2846.

57. C. D. Nenitzescu, I. Necsoiu, A. Glatz, and M. Zalman,
    Chem. Ber., $\underset{\sim}{92}$, (1959) 10.

58. J. E. Douglass and R. M. Roberts, J. Org. Chem., 28,
    (1963) 1229.

59. D. Farcasiu, Rev. Chim. (Bucharest), 10, (1965) 457.

60. A. Streitwieser, and L. Reif, J. Am. Chem. Soc., 86,
    (1964) 1988.

61. A. A. Khalaf, Ph.D. Dissertation, University of Texas,
    1965.

62. J. Shabtai, E. Lewicki, and H. Pines, J. Org. Chem., 27,
    (1962) 2618.

63. T. L. Gibson, Ph.D. Dissertation, University of Texas,
    1972.

64. R. M. Roberts, and T. L. Gibson, J. Am. Chem. Soc., $\underset{\sim}{93}$,
    (1971) 7340.

65. M. B. Abdel-Baset, Ph.D. Dissertation, University of
    Texas, 1973.

66. R. M. Roberts, T. L. Gibson, and M. B. Abdel-Baset, J.
    Org. Chem., $\underset{\sim}{42}$, (1977) 3018.

67. R. M. Roberts, Y. W. Han, C. H. Schmid, and D. A. Davis,
    J. Am. Chem. Soc., $\underset{\sim}{81}$, (1959) 640.

68. C. C. Lee, M. C. Hamblin, and J. F. Uthe, J. Org. Chem.,
    $\underset{\sim}{42}$, (1964) 1771.

69. R. L. Burwell, Jr. and A. D. Shields, J. Am. Chem. Soc.,
    $\underset{\sim}{77}$, (1955) 2766.

70. R. M. Roberts, S. E. McGuire, and J. R. Baker, J. Org.
    Chem., $\underset{\sim}{41}$, (1976) 659.

71. A. C. Olson, Ind. Eng. Chem., $\underset{\sim}{52}$, (1960) 833.

72. S. H. Sharman, J. Am. Chem. Soc., 84, (1962) 2945.

73. R. M. Roberts, G. P. Anderson, and N. L. Doss, J. Org.
    Chem., 33, (1968) 4259.

74. H. Steinberg and F. L. J. Sixma, Rec. Trav. Chim.
    Pays-Bas 81, (1962) 185.

75. E. Unseren and A. P. Wolf, J. Org. Chem., 27, (1962)
    1509.

76. G. G. Moore and A. P. Wolf, ibid., 31, (1966) 1106.

77. R. M. Roberts and S. Roengsumran, unpublished results.

CHAPTER 4

AROMATIC CATIONIC REARRANGEMENTS

D.L.H. WILLIAMS and E. BUNCEL

Department of Chemistry, University of Durham, Durham, England

and

Department of Chemistry, Queen's University, Kingston, Ontario, Canada

I   INTRODUCTION

Isotopes have played a major role in reaction mechanism studies of aromatic rearrangements. Important information regarding the inter- or intra-molecularity of the rearrangement can sometimes be obtained (e.g. in the rearrangement of phenylhydroxylamine), by noting in the product the pick up, or otherwise, of an isotope added to the reaction mixture in a form likely to exchange rapidly with the migrating fragment. Experiments with reactants specifically labelled in one position have in some cases (e.g. the Claisen rearrangement of allyl phenyl ethers) enabled the course of the reaction to be established. Rearrangements in diazonium ions can only be detected by specifically labelling one of the nitrogen atoms. Primary kinetic isotope effects have been widely used to pin point the rate limiting stage (e.g. in the benzidine rearrangement).

A common feature of all the reactions discussed in this chapter is that they are all initiated by mineral acid or Lewis acid catalysts, and generally involve, as reactive intermediates, cationic species where the positive charge is located on a heteroatom of a functional group, or on a ring carbon atom. The rearrangements discussed are summarised by the following equations:

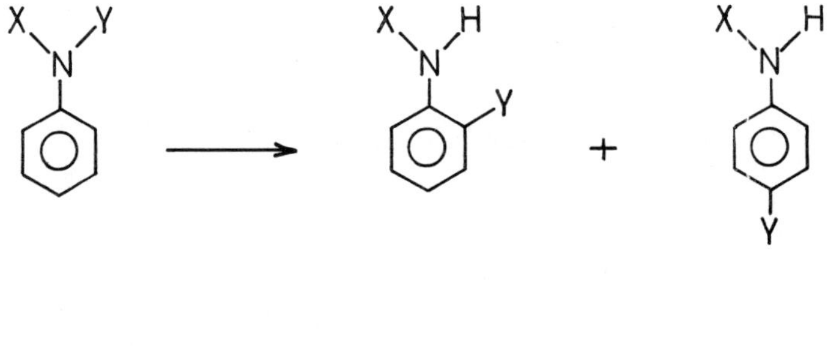

$$\text{(1)}$$

X = H, alkyl, aryl or acyl

Y = NO_2 (nitramine rearrangement),

NO (Fischer-Hepp rearrangement),

OH (phenylhydroxylamine rearrangement),

Cl (Orton rearrangement),

SO_3H (phenylsulphamic acid rearrangement).

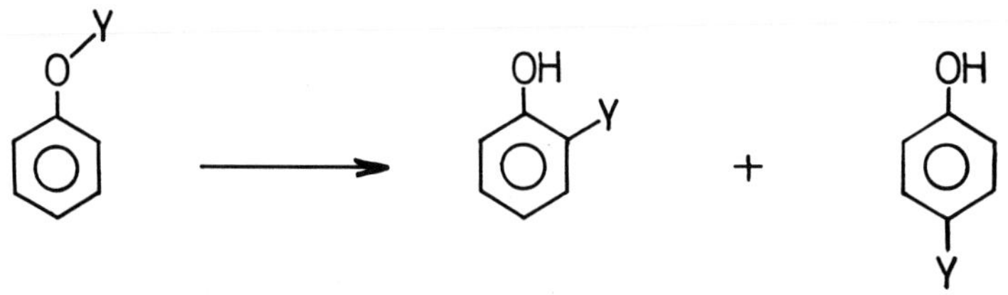

$$(2)$$

Y = acyl (Fries rearrangement),

alkyl, allyl (Claisen rearrangement).

Rearrangement of polyaromatics e.g.

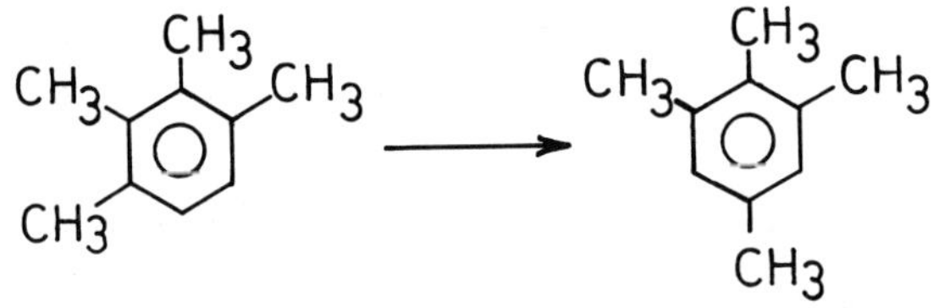

$$(3)$$

The Wallach rearrangement:

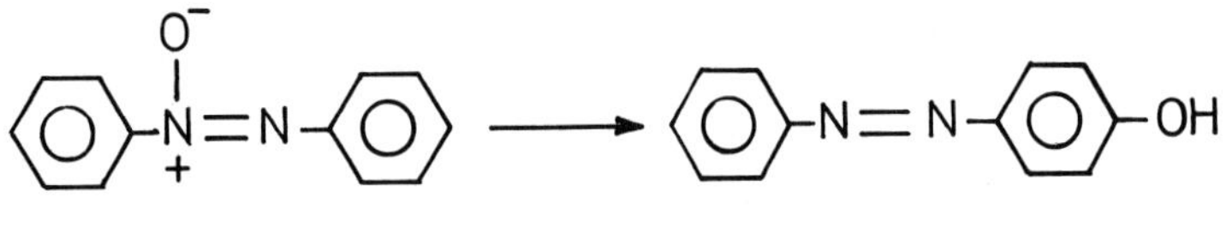

$$(4)$$

The benzidine rearrangement:

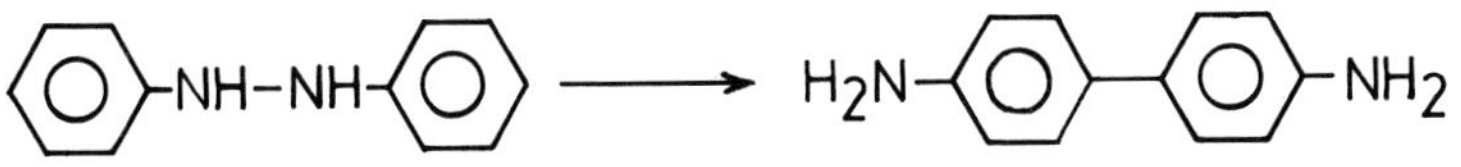

$$(5)$$

The diazonium ion rearrangement:

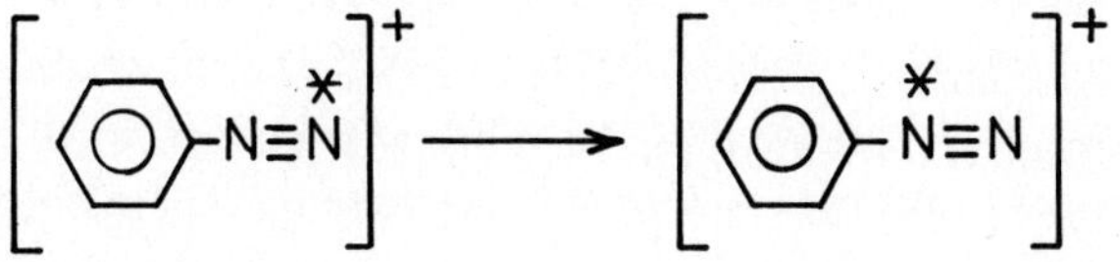

$$\left[\text{⬡—N≡\overset{*}{N}} \right]^{+} \longrightarrow \left[\text{⬡—\overset{*}{N}≡N} \right]^{+} \qquad (6)$$

II   THE NITRAMINE REARRANGEMENT

The reaction of N-nitroaniline derivatives (1) in acid media generally gives a mixture of the ortho- and para-C-nitroanilines (2 and 3), eqn. 7.   Reaction occurs in a wide

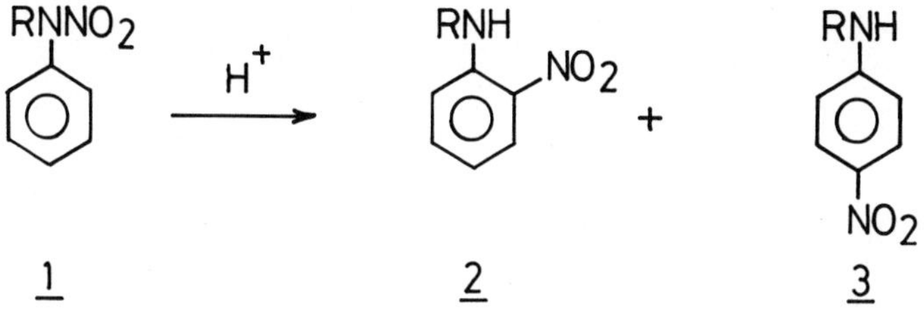

$$\underset{\underline{1}}{\overset{RNNO_2}{⬡}} \quad \xrightarrow{H^+} \quad \underset{\underline{2}}{\overset{RNH}{⬡}} NO_2 \quad + \quad \underset{\underline{3}}{\overset{RNH}{\underset{NO_2}{⬡}}} \qquad (7)$$

variety of solvents and with a number of different catalysing acid species.   There is no suggestion of an apparent specificity of hydrogen chloride as in the corresponding reaction of the N-nitroso compounds.   The reaction has been known since its discovery by Bamberger[1] in 1893 and is quite general for aniline and naphthylamine derivatives.   This rearrangement has been the subject of much mechanistic work over the last twenty five years or so, where the use of isotopes has played an important part.   In spite of this effort, however, there is currently no general agreement on a single mechanism for the reaction.

Usually the ortho-isomer predominates in the rearrangement products (often taken as a guide to an intramolecular reaction), but the ortho:para ratio varies quite significantly with the solvent and also with the acid concentration.

A mechanism involving fission of the N-N bond to give the free nitronium ion followed by a C-nitration was ruled out[2,3]

by the observations of quite different isomer proportions
from the rearrangement and direct nitration of the aniline
reactions.   Transfer (or cross) nitrations have been noted[3]
but in general these do not give any definitive evidence of
mechanism as the nitramine (or more probably its protonated
form) could itself act as a direct nitrating agent, without the
formation of the nitronium ion or a similar species.

Several attempts have been made to establish the intra-
or inter-molecularity of the rearrangement, using isotopic
labelling experiments.   Most of these conclude that the
reaction is intramolecular.   Reaction of $\underline{1}$(R=H) in sulphuric
acid solution in the presence of either [^{15}N] potassium
nitrate, [^{15}N] nitric acid[4,5] or [^{15}N] potassium nitrite
yielded both $\underline{\text{ortho}}$- and $\underline{\text{para}}$-nitroanilines without ^{15}N
enrichment.   Similarly[6], $\underline{\text{N}}$-methyl-$\underline{\text{N}}$-nitro-1-naphthylamine and
$\underline{\text{N}}$-nitro-1-naphthylamine gave rearrangement products free from
excess ^{15}N when the reaction was carried out (over a wide
range of acidity) in the presence of added ^{15}NO$_2^-$ and ^{15}NO$_3^-$.
Labelling of the substrate with ^{15}N has also been used to
establish the same mechanistic point.[7]   The concurrent
rearrangements of N-[^{15}N]nitroaniline ($\underline{1}$) and $\underline{\text{N}}$-nitro-$\underline{\text{para}}$-
toluidine ($\underline{4}$) gave 2-nitro-$\underline{\text{para}}$-toluidine ($\underline{5}$) without ^{15}N
enrichment (eqn. 8).   All of these results argue very strongly

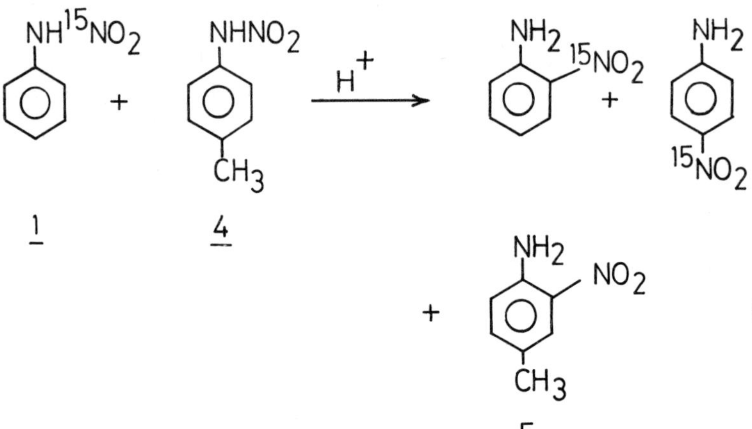

$\underline{1}$ $\underline{4}$

$\underline{5}$

(8)

in favour of an intramolecular mechanism for the rearrangement, since it is reasonable to expect that any scission fragment from the nitramine becoming free in solution would exchange ^{15}N label with either HNO_2 or HNO_3. Further, the experiment with the mixed nitramines[7] eliminates, in this case, any mechanism involving direct transfer of the nitro group. However, another report[8] indicates partial uptake of ^{15}N in the rearrangement product, para-nitro-N-methylaniline, when the simultaneous reactions of N-methyl-N-nitroaniline and para-fluoro-N-[^{15}N]nitro-N-methylaniline were carried out. This was interpreted by the authors as evidence for two reaction pathways, one intramolecular and one intermolecular, although other explanations are possible, such as a degree of reversible disproportionation (aided by the N-methyl group) accompanying rearrangement. The case for an intermolecular component would have been strengthened if it had been shown that the para-fluoro nitramine retained its full ^{15}N enrichment during the reaction.

As a result of the ^{15}N studies and rate measurements of the rearrangement reaction, two separate mechanistic schemes have been proposed. Scheme 1 is that advocated by Banthorpe and co-workers[5,6] and involves a completely intramolecular transfer of the NO_2 group to both ortho and para positions in the aromatic ring.

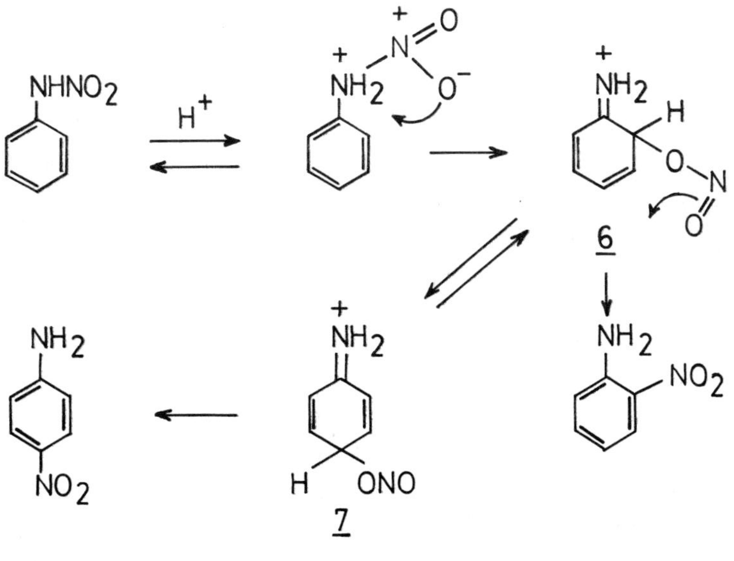

SCHEME 1

This is achieved by the formation of a reactive σ-bonded intermediate 6 (via a cyclic transition state) which can break down directly to give the ortho-nitro product or form 7 (via another cyclic transition state) which then collapses to give the para-product. Cyclic transition states of this type have been advocated in other rearrangements e.g. in the Claisen rearrangement of allyl aryl ethers.

White and co-workers[8,9] prefer a caged-radical mechanism outlined in Scheme 2. Here the protonated nitramine forms a radical cation-radical pair (8) (retained within the solvent

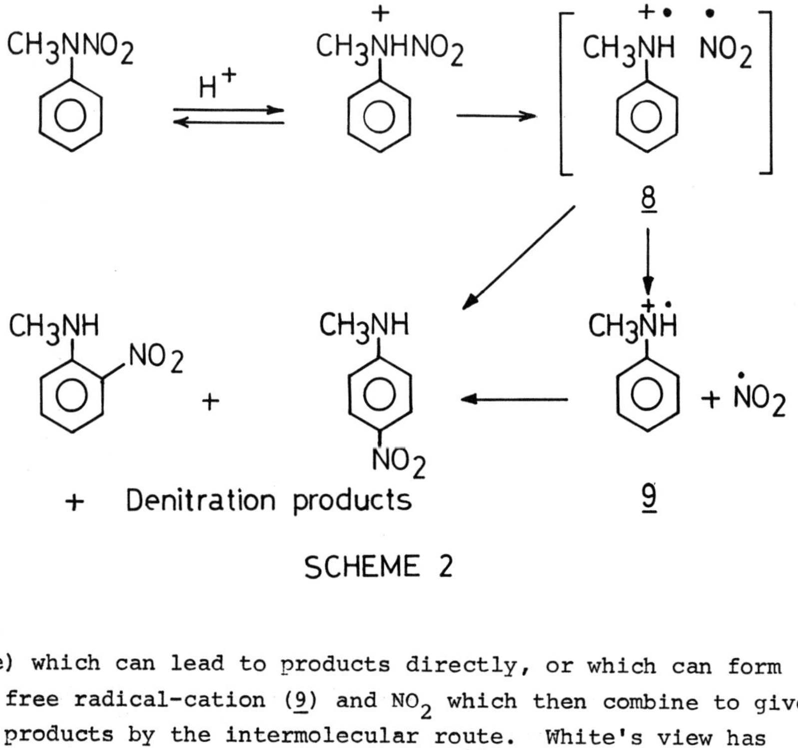

SCHEME 2

cage) which can lead to products directly, or which can form the free radical-cation (9) and NO_2 which then combine to give the products by the intermolecular route. White's view has been presented in detail in a review.[10]

Both groups of workers agree that the first stage is the rapid equilibrium formation of the protonated form of the nitroamine, probably as the N-protonated species, although a suggestion has been made[11] that reaction might occur via an O-protonated species. The first-order rate constant is dependent upon the Hammett acidity function and there is a

large solvent isotope effect (k_{D_2O}: k_{H_2O} 3.3 typically[5] for
N-nitroaniline in D_2SO_4/D_2O) characteristic of reaction via a
low concentration of a rapidly formed conjugate acid.   There
is agreement also that the final proton-transfer to the solvent
is not rate-limiting, as is generally found in nitration
reactions.   This was shown by the absence of a primary kinetic
isotope effect (k_H:k_D~1) when ring deuterated reactants were
compared with the ^{1}H substrates.[5,6]   Product isotope effects
have however been reported.[5]   The ortho:para ratio of the
product nitroanilines was measured (a) for the 2,3,5,6-
tetradeuterated, and (b) for the 2,4,6-trideuterated reactant
nitramine.   The results which are presented in Table 1 are
compared with those found for the isotopically normal substrates.

TABLE 1
Variation of the ortho : para ratio upon ring deuteration

N-Nitroaniline	Solvent	Ortho : para ratio
Normal	75% H_2SO_4	19
2,3,5,6-D_4	75% H_2SO_4	19
Normal	37% $HClO_4$	2.6
2,3,5,6-D_4	37% $HClO_4$	2.1
Normal	50% $HClO_4$	3.0
2,4,6-D_3	50% $HClO_4$	4.6
Normal	26% $HClO_4$	2.3
2,4,6-D_3	26% $HClO_4$	4.9

For (a) where the ortho- (but not the para-) positions are
isotopically substituted, the change had no effect on the
ortho : para ratio at high acidity (75% H_2SO_4), but changed
the ratio from 2.6 to 2.1 at lower acidity (37% $HClO_4$).   With
both ortho- and para- positions labelled as in (b), the ortho:
para product ratio increased from 3.0 to 4.6 in one case (in
50% $HClO_4$) and from 2.3 to 4.9 at a lower acidity (26% $HClO_4$).
These changes, which were well outside the experimental
measurement error, were considered to be significant and

interpreted as a product isotope effect arising from a direct
competition (after the rate-limiting step) between protein loss
from a C-protonated precursor of the ortho-product (like 6) and
its transformation to an intermediate (7) which leads to the
para-isomer. Similar effects were observed[12] in the reactions
of N-nitro-1-naphthylamine and its 2,4-dideutero derivative.
These results can readily be accommodated by the cyclic
transition state theory (Scheme 1) but not by the caged
radical mechanism of Scheme 2. However, White and co-workers[13]
found no such product isotope effect in the reaction of the
2,6-dideutero derivative of N-methyl-N-nitroaniline in 0.5M
hydrochloric acid at 40°C. This observation was taken as
support for the caged-radical mechanism. Tests for radicals
during rearrangement, by e.s.r. and by the intiation of poly-
merisation, however, have proved negative.[6]

Before leaving this work it is perhaps worth noting that
White et al.[9,13] have made much use of the ^{14}C isotope in
isotope dilution analysis to establish the yields of the
various products (ortho- and para-nitro-N-methylaniline and
N-methylaniline) from the reaction of N-methyl-N-nitroaniline.
This was carried out using ^{14}C-labelled reactant prepared from
uniformly labelled ^{14}C- aniline, by dilution with inactive
products, and by degradation to carbon dioxide.

Banthorpe and Winter[14] have examined the acid-catalysed
rearrangement of N-methyl-N-nitro-9-aminoanthracene (10),
originally as an experiment designed to distinguish clearly
between the cyclic-transition state mechanism and the radical-
cage mechanism. In sulphuric acid (3.5M) 10-nitroanthrone
(13) was obtained in high yield. Again ^{15}N studies showed
that the rearrangement was an intramolecular process, with a
possible maximum 7% intermolecular component. However, this
nitramine substrate showed a most unusual solvent isotope
effect of k_{D_2O} : k_{H_2O} of 0.84. This compares with the values
of 2.5 to 3.3 found generally for N-nitroaniline, N-nitro-1-
naphthylamine and their N-methyl derivatives.[5,12,13] This was
interpreted as involving a rate-limiting proton-transfer to
the aromatic ring to form the intermediate 11 in Scheme 3.
Rearrangement of 11 then was considered to occur intramolecularly to give 12 from which the product is readily formed

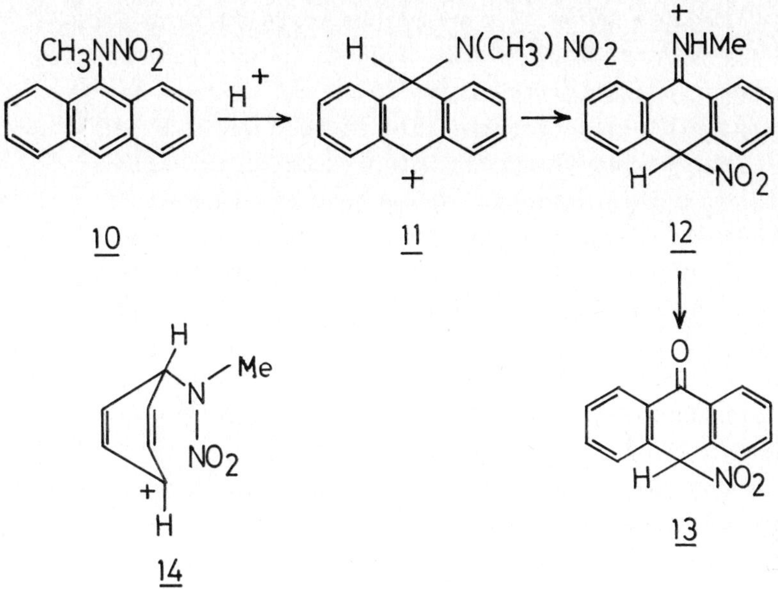

SCHEME 3

by proton loss and methylamine loss. The actual rearrangement step 11→12 could conceivably go via a direct reaction involving the bent boat form transition state 14[15] a tight ion-pair intermediate[5] or a radical-ion cage.[9]

Whilst there is no one unified mechanism as yet agreed for the nitramine rearrangement, it is worth noting that for many of the experiments carried out, the reaction conditions (substrate, acid concentration, temperature and solvent) have in many cases been significantly different. It is perhaps too much to expect one mechanism to prevail.

The rearrangement is well known in the heterocyclic field; an intramolecular pathway is indicated from ^{15}N studies for the pyridine series.[16]

One further interesting point has recently been established. Bamberger[17] originally believed that C-nitration of aromatic amines occurred first by N-nitration, followed by rearrangement. The fact that markedly different product isomer ratios were found for rearrangement and nitration was taken as evidence against this idea.[2,3] It is now known that nitration of aromatic amines does in many cases proceed <u>via</u> the protonated form of amine,[18] and so the comparison is not valid. Using a much weaker base, 2,3-dinitroaniline,[19] which undergoes nitration <u>via</u> the free amine form, the nitroamine has been isolated as an intermediate in good yield in the nitration process, and very similar product distributions obtained from both reactions. These results are also consistent with a mechanism involving de-nitration of the nitroamine, forming free nitronium ions and the free amine, which can then undergo a conventional <u>C</u>-nitration. This explanation was eliminated by demonstrating[19] the intra-molecular nature of the rearrangement using [^{15}N]HNO$_3$ added to the reaction mixture; no significant pick up of ^{15}N was observed.

It is thus appropriate to end this section by confirming one of the early ideas of Bamberger who did so much pioneering work with the nitramine and other aromatic rearrangements.

III   REARRANGEMENT OF <u>N</u>-NITROSO AMINES (FISCHER-HEPP)

The acid-catalysed rearrangement of <u>N</u>-nitroso-<u>N</u>-alkyl-aniline derivatives (<u>15</u>) was established by the early work of Fischer and Hepp[20] in the last century. The rearrangement product is always the <u>para</u>-nitroso isomer (<u>16</u>) (except in the case of the 2-naphthylamine derivative), which is sometimes accompanied by varying amounts of the product of denitrosation, the corresponding secondary amine (<u>17</u>), eqn. 9. The reaction is quite general and gives high yields of <u>16</u> when carried out

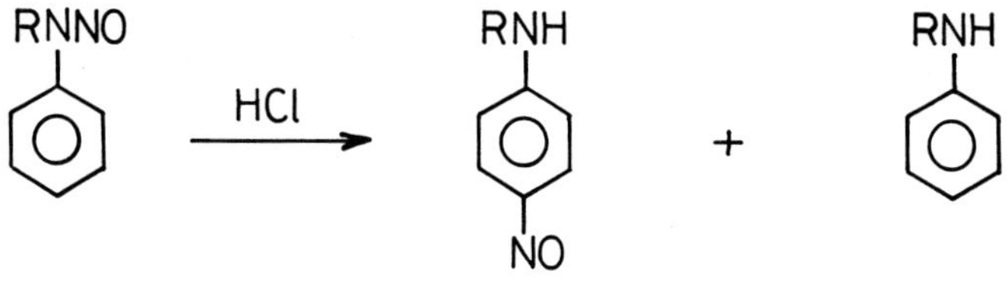

$$\text{15} \xrightarrow{\text{HCl}} \text{16} + \text{17} \tag{9}$$

with hydrogen chloride in dry ethanol or ether solvent. In
aqueous acid solutions there is a tendency for more of 17 to be
formed. The observation of the apparent specificity of
hydrogen chloride (cf.the Orton rearrangement of N-chloroacetanilide), together with the results[21,22], of a number of
product analyses where nitrosation of added substances (such as
reactive amines and alkenes) occurred, have been taken to
support the suggestion of Houben,[22] that the reaction proceeds
by way of a chloride ion-catalysed denitrosation to give the
free secondary amine and nitrosyl chloride (reversibly), which
then effect a conventional electrophilic C-nitrosation at the
para-position of the aromatic ring (see Scheme 4). This out-
line mechanism has subsequently been widely accepted and
reported in standard texts and review articles, although some
authors have pointed out that the detailed kinetic evidence for
such a mechanism is lacking,[23] and Dewar[24] wrote that these
early experiments give in fact no definite evidence whatsoever
concerning the mechanism. In some cases the "evidence" was
accepted uncritically; for example, the failure of meta-nitro-

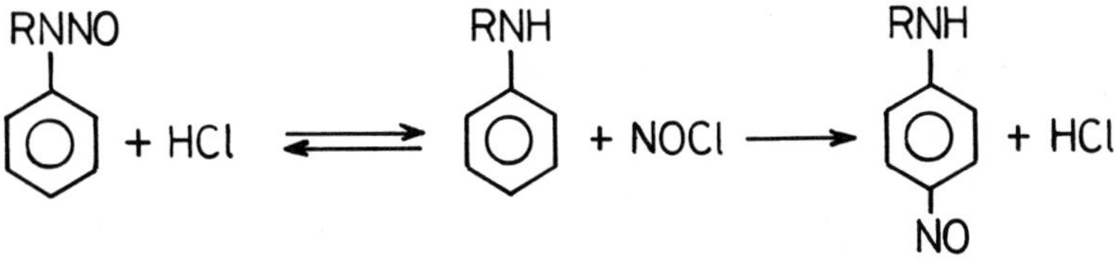

SCHEME 4

N-methyl-N-nitrosoaniline to rearrange in the presence of
excess urea[25] (a well-known trap for free nitrous acid/nitrosyl
chloride) was taken as support for such a scheme, although it
was not shown that rearrangement occurred in the absence of
urea. Later work[26] showed that the meta-nitro substituent so
deactivates the para-position to electrophilic nitrosation
that rearrangement of this nitrosamine has never been
accomplished.

    More recently a detailed kinetic study of the reaction in
aqueous acid solutions has been carried out. This shows[27]
that rearrangement and denitrosation occur by concurrent

separate pathways, from a common intermediate, the protonated
form of the nitrosamine (18). Such a mechanism is outlined in
Scheme 5.

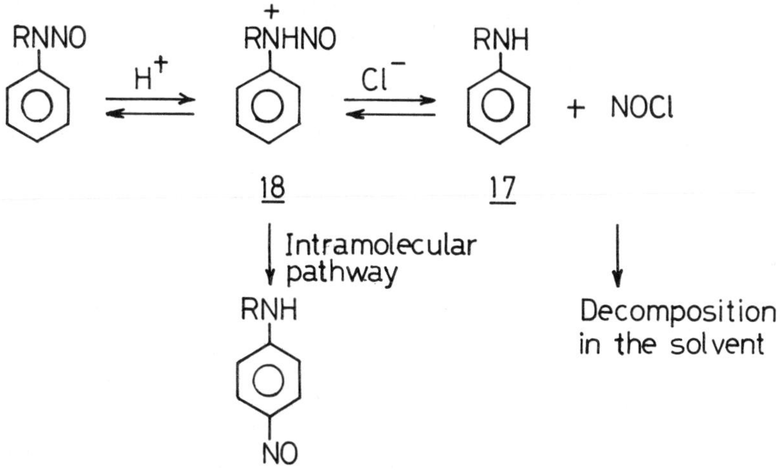

SCHEME 5

This scheme is not only consistent with the recent kinetic
findings (which are not reproduced here), but also accounts
for the early cross-nitrosation experiments. It was possible
to examine denitrosation and rearrangement separately as limiting forms by a suitable choice of reaction conditions, i.e.
denitrosation only, by addition of a nitrite trap (urea,
sulphamic acid etc.) in sufficient concentration, and
rearrangement alone, by adding a sufficient excess of the
secondary amine 17 to suppress denitrosation completely.
Several kinetic and product analysis features[27,28] enabled
Scheme 4 to be discarded in favour of Scheme 5.

Initially it was believed that the question of inter-
versus intramolecularity of this rearrangement could be
settled by the pick-up or otherwise of an isotopic label,
added to the reaction mixture. Indeed when the rearrangement
was carried out in the presence of ca. 30% enriched [^{15}N]NaNO$_2$,
the product para-nitroso isomer was almost fully labelled.
However, it was established that rapid equilibration of the
label occurred between the added nitrite and the nitrosamine,

as is shown in Table 2.

TABLE 2

^{15}N analysis of the product of rearrangement and recovered reactant

% conversion	Ratio 137:136 as a % from the mass spectra of <u>para</u>-nitroso-<u>N</u>-methylaniline*	
	(a) from recovered reactant	(b) from product
6	20.9	20.7
20	21.1	21.1
45	-	21.3
75	-	21.1

* 21% corresponds to complete equilibration.

In these experiments, it was found to be more convenient to obtain mass spectra of samples of the <u>para</u>-nitroso isomer throughout (because of the difficulty experienced in obtaining a reasonable parent ion peak in the spectrum of the nitrosamine), so that the recovered reactant, which was separated from the product by column chromatography, was converted to the <u>para</u>-nitroso isomer in a separate reaction using hydrogen chloride in dry ethanol. Clearly no conclusions regarding the nature of the rearrangement can be drawn from these experiments since the reactant equilibrates ^{15}N with added nitrite before a significant amount of rearrangement has occurred.

Isotopic substitutions have, however, enabled certain features of the mechanism to be established. Both rearrangement and denitrosation reactions[29,30] gave a kinetic solvent isotope effect of $(k)_{D_2O}:(k)_{H_2O}$ of <u>ca</u>. 2.5. This suggests (in both cases) that a rapid reversible protonation of the nitrosamine occurs, which is followed by some rate-limiting step.

Information regarding the final proton-loss from the <u>para</u>- position of the aromatic ring has been obtained by noting the value of the kinetic isotope effect $k_H:k_D$ when deuterium is substituted for hydrogen in the aromatic ring. A value of 2.4 was obtained[26] for the rearrangement of <u>N</u>-methyl-<u>N</u>-nitrosoaniline under conditions when denitrosation was suppressed (i.e. in the presence of an excess of ring-deuterated N-methylaniline). Similar values (2.6 and 2.7) were obtained at two different acidities for the rearrangement of meta-methoxy-N-methyl-N-nitrosoaniline, which is virtually quantitative under these conditions. Clearly, for both substrates, the final proton-loss is at least in part rate-limiting. Although kinetic isotope effects are not common in electrophilic aromatic substitution (whether intra- or intermolecular), effects of this magnitude have been observed in the C-nitrosation of phenols[31] and other aromatic systems;[32] indeed, this seems to be a common feature of aromatic C-nitrosation reactions.

It has been possible to distinguish between the inter- and intramolecular mechanisms (Schemes 4 and 5, respectively) by a detailed product analysis study involving addition of 'traps' for free nitrous acid (or nitrosyl halide).[28] Above a certain threshold concentration of the trap (such as sulphamic acid or sodium azide), the reaction rate constant became independent of the concentration and nature of the trap and the yield of rearrangement product attained a non-zero limiting value. The results are summarised in Table 3 for the reaction of N-methyl-N-nitrosoaniline in 2.75M sulphuric acid. Scheme

Table 3

Yield of rearrangement and rate constants for a variety of traps

Trap	[Trap]/M	$10^4 k_o/s^{-1}$	% Rearrangement
HN_3	6.5×10^{-4}	0.65	21
HN_3	16.3×10^{-4}	0.67	21
NH_2SO_3H	3.1×10^{-3}	0.65	21
NH_2SO_3H	7.8×10^{-3}	0.64	21
$CO(NH_2)_2$	0.10	0.62	21
NH_2OH	2.6×10^{-3}	0.62	20
NH_2NH_2	1.6×10^{-3}	0.66	20

4 requires a direct competition for the free nitrous acid (or nitrosyl chloride) between the secondary amine produced and the trap (sulphamic acid etc.). This must lead to the reduc-

tion of yield of rearrangement product towards zero as the trap concentration and/or reactivity is increased.

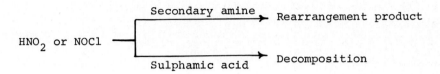

This is contrary to the observation that, for reaction in 2.75$\underline{M}$ sulphuric acid, a constant ratio of 21% rearrangement to 79% denitrosation is found for a wide variety of trap species. Scheme 5 on the other hand, is perfectly compatible with this result, in fact demands a constant product ratio at high trap

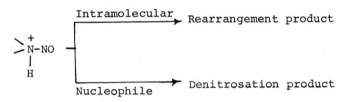

concentration where denitrosation is now irreversible and we have the situation of two parallel reactions. A more reactive nucleophile diverts the reaction more towards denitrosation and often no rearrangement product is formed. The constant k_o value found over the range is also consistent with such a mechanistic scheme.

A detailed kinetic analysis[27] also enables a distinction between the two mechanisms to be made on the basis of chloride ion (or any other suitable nucleophile) catalysis. Full details will not be presented here, but it is clear that the intramolecular mechanism (Scheme 5) predicts no chloride ion catalysis at constant acidity, in the presence of an excess of added secondary amine product ($\underline{N}$-methylaniline in this case). Under these conditions rearrangement is virtually quantitative and is found to be not subject to chloride or bromide ion catalysis (Table 4). Under comparable conditions the rate constant for denitrosation is increased by a factor of 2.6 by

Table 4

Rate constants for rearrangement in the presence of halide ion

[N-Methylaniline]added	[Halide] added	$10^4 k_o (s^{-1})$
4×10^{-3} M	O	1.75
4×10^{-3} M	0.24M NaCl	1.79
4×10^{-3} M	0.10M NaBr	1.77

0.24M NaCl and by 4.6 by 0.10M NaBr. The rate equation derived from Scheme 4 for the intermolecular mechanism predicts that chloride ion catalysis should always be present, contrary to the observed results.

The mechanistic conclusions to be drawn from these and the kinetic experiments regarding this rearrangement are summarised in Scheme 6. It is probable that intermediates 18

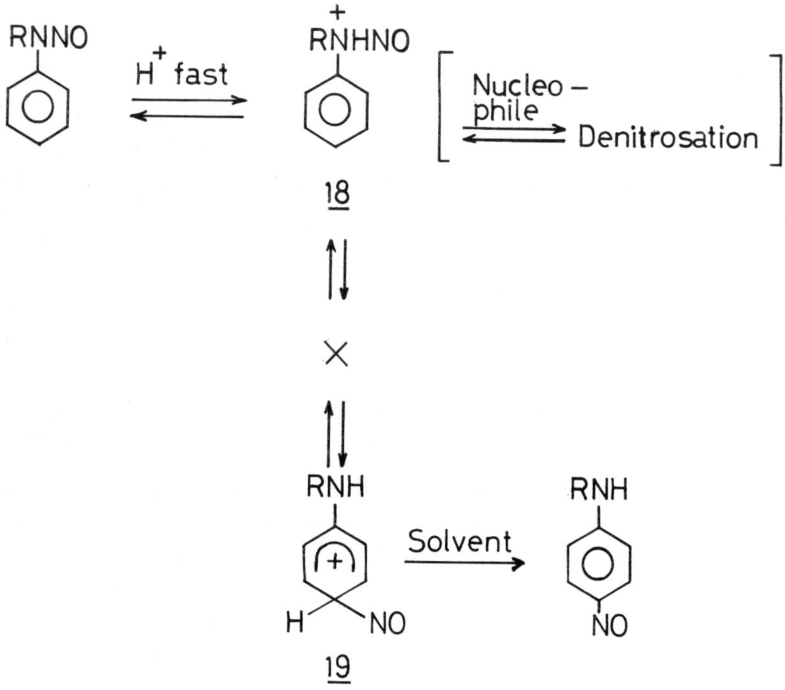

SCHEME 6

and 19 lie on the path towards rearrangement, although it is

unlikely that the change 18 → 19 can be effected directly.
We have therefore written an intermediate X whose nature must,
at this stage, be a matter for speculation. Conceivably it
could be a π-complex, or related species 20, or the bent ring-
protonated species 21. Kinetic substituent effects[26] show

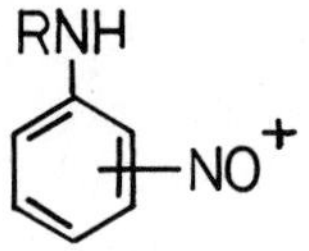

20 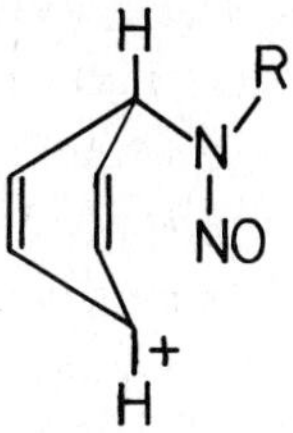   21

that the rearrangement is one of overall _electrophilic_ nitrosation. This would seem to rule out nitrosation _via_ nitric
oxide (a poor nitrosating agent), either free or in the form
of an intermediate cation radical-radical pair formed directly
from 18. Vrba and Allan[33] have suggested an intermediate of
the type 21 for the phenylsulphamic acid rearrangement. Later
this was extended[34] to include the Fischer-Hepp, nitramine and
benzidine rearrangements (_vide infra_). One of the difficulties
in accepting such a reaction pathway is that there is no
evidence for the C-1 protonation required. Here the matter
rests at present; it is not easy to envisage additional
experiments which would clarify the mechanism further and
also account for the exclusive formation of the _para_- (and no
ortho-) isomer.

IV   REARRANGEMENT OF PHENYLHYDROXYLAMINES

In 1M sulphuric acid at room termperature,
phenylhydroxylamine (22) is readily converted to _para_-amino
phenol (23), eqn. 10. Much of the early work on this
reaction was carried out by Bamberger.[35] All the product

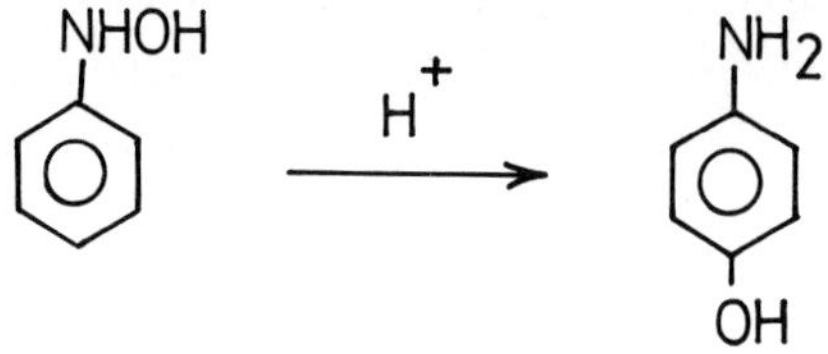

$$NHOH \quad \xrightarrow{H^+} \quad NH_2 \qquad (10)$$

22   23

analysis work (mainly work by Bamberger) points to an inter-molecular mechanism. For example, in the presence of nucleophiles other than the solvent water, other products are formed viz, phenetidines from ethanol solutions and chloroanilines from hydrochloric acid solution. This view was confirmed[36] by the observation that full uptake of ^{18}O occurred in the product, when reaction was carried out in ^{18}O enriched water solvent. More recently a more comprehensive examination of this ^{18}O uptake was undertaken by Kukhtenko.[37] He confirmed that full uptake occurs in the product para-amino-phenol and no incorporation of ^{18}O occurs in the starting material. This is an important point to establish, as it eliminates an intramolecular mechanism which takes place concurrently with reversible disproportionation of the reactant (or its protonated form), which is indeed the situation in the case of the Fischer-Hepp rearrangement of N-nitrosamines. It was shown further[37] that the product was not formed via the intermediate formation of para-aminophenyl sulphate. This was achieved by effecting the rearrangement in the presence of $[^{18}O]H_2S_4$ and observing no uptake of ^{18}O in the para-amino-phenol product.

Kinetic studies[38] have shown that the reaction is acid catalysed at low acid concentration, where the rate constant is proportional to $[H^+]$, whilst at higher acid concentrations the rate becomes independent of the acidity. There was no evidence of any second-order acidity dependence (as in the benzidine rearrangement) and the reaction rate (though not the product distribution) was unaffected by the presence of chloride ions. The modern view of the reaction mechanism, first set out by Heller, Hughes and Ingold[38], is essentially

the same as that held by Bamberger in the 1920's and is given
in Scheme 7. Loss of a water molecule from the protonated form
of the reactant 24 leads to the intermediate 25 which undergoes

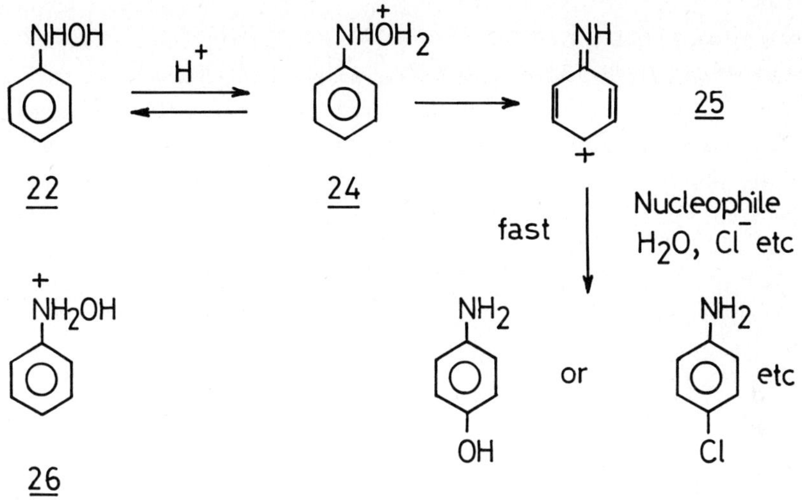

SCHEME 7

rapid attack by nucleophilic species present. Concurrent
attack by the nucleophile and loss of water molecule from 24
is ruled out by the absence of a kinetic dependence on e.g.
Cl^-. An intermediate similar to 25 can be written with the
positive charge located at the ortho position, which would
give the ortho- product, observed in some cases together with
the para-product.

There are of course two possible sites of protonation in
the parent hydroxylamine, at O-, and N- to form 24 and 26,
respectively. It is to be expected that the nitrogen atom is
a more basic site than the oxygen atom, whereas it is easier
to visualise reaction via the O-protonated intermediate.

Very recently a full mechanistic study has been carried
out[39] (as yet unpublished) where use is made of isotopic
substitution in the solvent. At low acid concentration (below
$5 \times 10^{-3} M$ H_2SO_4) the observed first-order rate constant for
the rearrangement (k_o) is greater in D_2SO_4/D_2O than in
H_2SO_4/H_2O - a fact which is consistent with reaction via a low

concentration of a rapidly and reversibly formed protonated
species. At higher acidities the solvent isotope effect
$(k_o)_{D_2O} : (k_o)_{H_2O}$ became inversed, with a value of 0.65 at 0.2$\underline{M}$
H_2SO_4, which is on the plateau of the k_o $\underline{vs}$ acidity plot.
Effects of similar magnitude were noted for two substituted
phenylhydroxylamines, 0.71 (at 0.3$\underline{M}$ H_2SO_4) for the $\underline{N}$-ethyl
substrate and 0.65 (at 0.25$\underline{M}$ H_2SO_4) for the $\underline{para}$-methyl
reactant. These results can be interpreted in terms of a rate-
limiting proton transfer. At these acidities it is probable
that the phenylhydroxylamines are present as the fully $\underline{N}$-
protonated species $\underline{26}$. Disproportionation can then occur by a
proton transfer reaction to form the $\underline{O}$- protonated species
(Scheme 8), which can readily lose the water molecule.

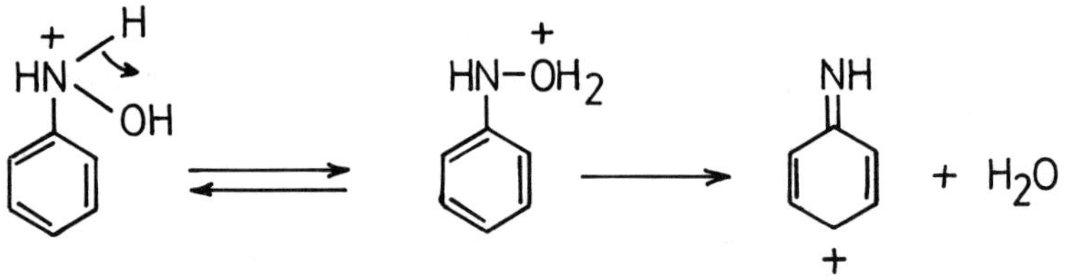

SCHEME 8

However, it is possible also to explain the observed solvent
isotope effects in terms of the equilibria between $\underline{22}$ and the
two protonated forms $\underline{24}$ and $\underline{26}$.

Loss of a water molecule from $\underline{26}$ could form the nitrenium
ion $\underline{27}$. Recently there has been much interest in the
chemistry of nitrenium ions, generated principally by solvolysis

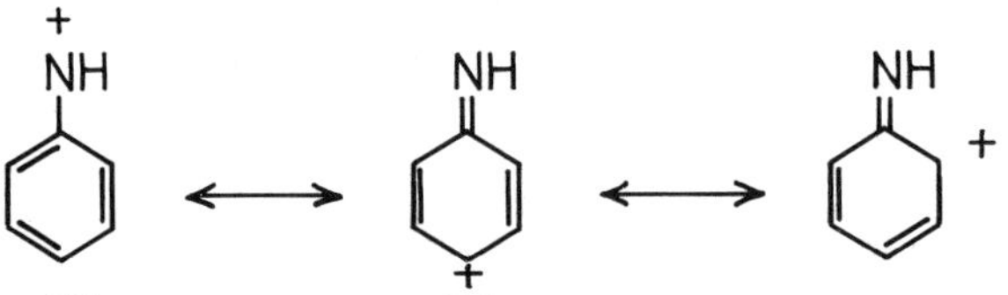

of $\underline{N}$-chloroaniline derivatives.[40] From phenylhydroxylamine,
however, it seems that the structure of the intermediate is

more accurately represented by <u>25</u> where the positive charge is
located at the <u>para</u> position of the aromatic ring, rather than
at the nitrogen atom. This is shown clearly by kinetic
substituent effects, where <u>N</u>-ethyl substitution gave a rate
enhancement of only 1.2 whereas a <u>para</u>-methyl substituent
increased the rate constant by a factor of over 10^2.

Rearrangement occurs also of <u>O</u>-substituted phenylhydroxylamines, generally to give the <u>ortho</u>-isomer. Two groups of
workers[41] used the ^{18}O-labelled <u>O</u>-(arenesulphonyl)
phenylhydroxylamine (<u>28</u>) and found the label attached to the
ring in the product <u>30</u>. This was taken as evidence for a
cyclic intramolecular mechanism involving an intermediate
transition state of the type <u>29</u>, similar to those suggested

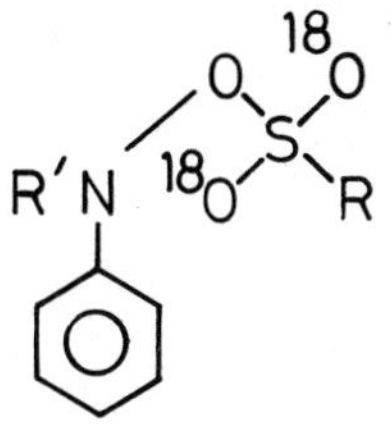

<u>28</u> 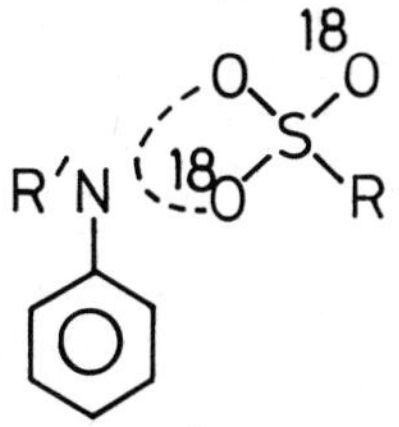   <u>29</u   <u>30</u>

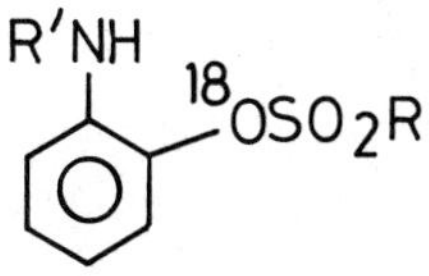

for the Claisen rearrangement of allyl phenyl ethers.

Later work[42] showed that, starting with the ^{18}O-labelled
compound <u>31</u>, 23% of the ^{18}O- ring bonded isomer <u>32</u> was formed,
the remaining 77% being in the sulphonyl group as in <u>33</u>. This
was interpreted in terms of a mechanism involving an ion pair
of the type <u>34</u> which has a very short lifetime, collapsing to
the product, <u>before</u> complete equilibration of the oxygen atoms
can occur. Substituent rate effects and solvent effects were
taken as support of this mechanism.

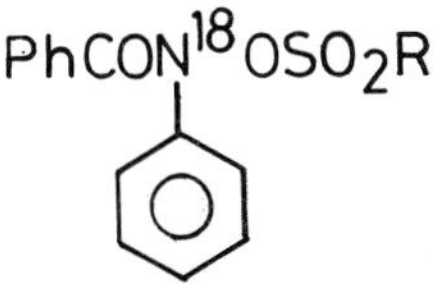

<u>31</u> 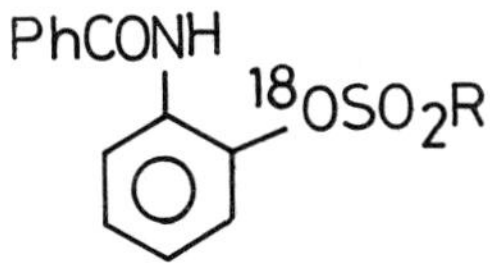   <u>32</u> 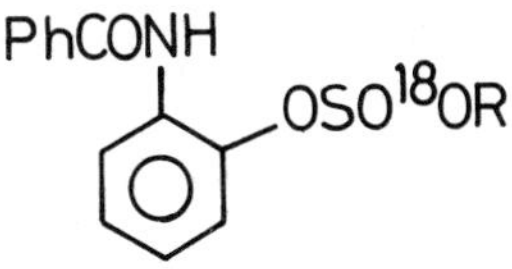   <u>33</u>

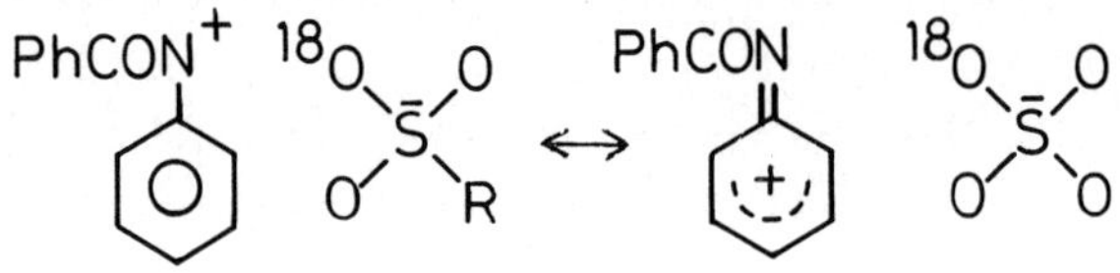

$$PhCON^{+} \quad {}^{18}O \underset{O}{\overset{O}{\underset{\parallel}{\overset{\parallel}{S}}}} R \quad \longleftrightarrow \quad PhCON \quad {}^{18}O \underset{O}{\overset{O}{\underset{\parallel}{\overset{\parallel}{S}}}} O$$

34

V REARRANGEMENT OF N-HALOANILIDES (ORTON REARRANGEMENT)

This rearrangement, typically of N-chloroacetanilide, to give para-chloroacetanilide, has been known for almost a hundred years,[43] and was examined kinetically by Orton and his co-workers between 1909 and 1928. These results were summarised by Hughes and Ingold[44] in 1952. Since that time relatively little mechanistic work has been done on this reaction in aqueous media although the reaction in other solvents (where the mechanistic picture is much less clear) has been studied on a number of occasions. A number of observations are consistent with an intermolecular mechanism for the rearrangement. Hydrochloric, hydrobromic and hydriodic acids act as catalysts for the change;[45] more reactive (towards electrophilic substitution) aromatic compounds, when added to the reaction, undergo halogenation and the ratio of isomers from the rearrangement is the same as that from the direct halogenation of the anilide.[46] Bromoanilides undergo an analogous reaction.[47] Chlorine can be aspirated from the reaction solution of a N-chloroanilide. The kinetic measurements, for reaction in hydrochloric acid, show that the rate is proportional to $[HCl]^2$ or to $[H^+][Cl^-]$. All of these facts support the mechanism (outlined in Scheme 9) which involve N-protonation of the anilide, a reversible de-chlorination brought about by nucleophilic attack by chloride ion, and a conventional electrophilic aromatic chlorination of the anilide formed by the chlorine produced.

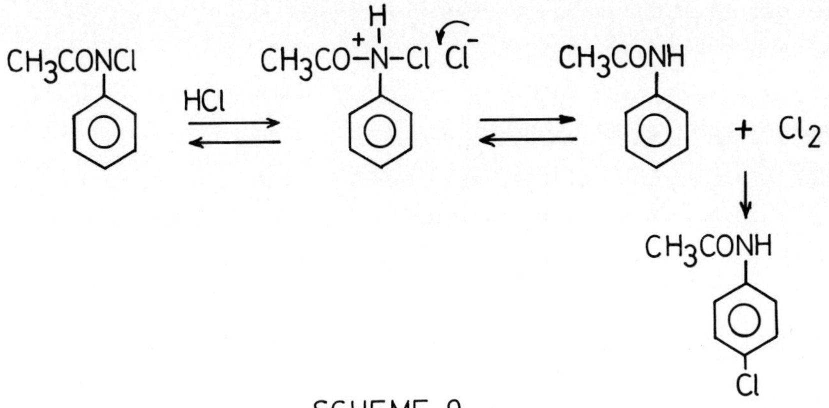

SCHEME 9

Little use has been made of isotopic substitutions in this case.
No data have been reported for reactions involving deuterated
acids, nor has there been a measurement of the kinetic isotope
effect upon ring-deuteration. It has been shown,[48] however,
using radioactive chlorine added as chloride ion, that full
uptake of radiochlorine occurs in the product para-chloroacetanilide, to the extent expected if full equilibration between
inorganic and organic chlorine has taken place. This result
is, of course, consistent with the reaction sequence outlined
in Scheme 9.

Rearrangement also occurs in aprotic solvents (such as
chlorobenzene) in the presence of catalysts such as acetic,
benzoic and picric acids.[49,50] Here the kinetics are more
complicated and no clear-cut mechanism has emerged. It has
been claimed that an intramolecular component of the reaction
exists. Reaction of N-bromoacetanilide in the presence of
[14]C-labelled acetanilide[51] gave radioactive para-bromoacet-
anilide, but not at the level expected from a fully intermolecular process. The results have been interpreted in terms
of a fast exchange process which occurs more rapidly than does
the rearrangement:

$$PhNBrAc + \overset{*}{P}hNHAc \rightleftharpoons PhNHAc + \overset{*}{P}hNBrAc \qquad (11)$$

It has been claimed[51] that experiments with the [82]Br isotope confirm that this exchange process occurs, although this work has not been published in full.

VI REARRANGEMENT OF ARYLSULPHAMIC ACIDS

Phenylsulphamic acid (35) when heated at around 100°C in fairly strong sulphuric acid is readily converted to sulphanilic acid (36):

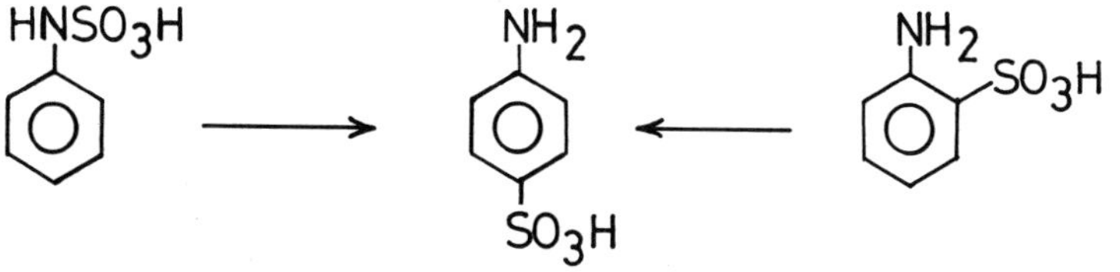

$$
\text{HNSO}_3\text{H} \quad \longrightarrow \quad \begin{array}{c}\text{NH}_2 \\ \\ \text{SO}_3\text{H}\end{array} \quad \longleftarrow \quad \begin{array}{c}\text{NH}_2\text{—SO}_3\text{H}\end{array} \quad (12)
$$

$$35 \qquad\qquad 36 \qquad\qquad 37$$

Other reaction conditions can also bring about this change. Various groups of workers have examined this reaction mechanistically, many using isotopic modifications, but no clear cut picture of the mechanism has emerged. Bamberger[52] in his early work on this reaction claimed that 35 first rearranged easily at low temperatures to orthanilic acid (37) by way of an intramolecular reaction, and the subsequent change, 37→36, took place at high termperatures. It is known that 37 will isomerise to 36 under these conditions. Later, however, Illuminati[53] was unable to recover any 37 from the reaction mixture, using Bamberger's stated reaction conditions; it seems that subsequently no one has been able to isolate any orthanilic acid from this reaction, so that Bamberger's result is now regarded as something of a mystery. Under conditions where 37 will not form 36 (dry dioxan containing sulphuric acid at 100°C), 35 was shown[53] to yield some 36. These reactions are, of course, closely related to the direct sulphonation of aniline. It was originally believed[52] that 35 was an intermediate in this reaction, i.e. that N-sulphonation first occurred, followed by rearrangement; this has never been unambiguously proved.

The ^{35}S isotope has been used by several workers in an

attempt to reveal the mechanism of the rearrangement.
Reaction of 35 (added as the potassium salt) in dioxan
containing [^{35}S]sulphuric acid gave^{54} sulphanilic acid (36)
with the full activity expected if the rearrangement were fully
intermolecular and the migrating species exchanged label
completely with the sulphuric acid present. Both sulphanilic
acid and orthanilic acid were recovered unchanged and inactive
when subjected to the rearrangement conditions. This
eliminates the possibility that orthanilic acid is an intermediate in the rearrangement process. These results were
interpreted in terms of a mechanism involving desulphonation of
the sulphamate to give aniline and sulphuric acid, which then
effected a ring C-sulphonation at the para-position (Scheme 10).
One drawback was that aniline itself could not be sulphonated

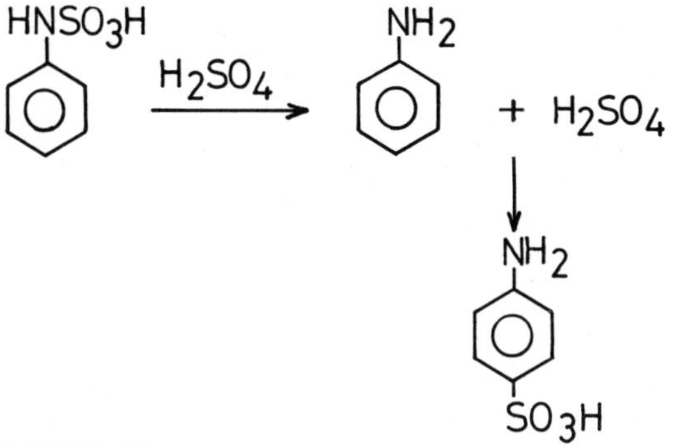

SCHEME 10

under these conditions. Kinetic measurements indicated that
the desulphonation step was rapid, with the C-sulphonation
presumably being rate-determining. About the same time, Vrba
and Allan^{55} found that phenylsulphamic acid undergoes a rapid
(in terms of the rate of rearrangement) exchange of label with
[^{35}S]sulphuric acid. They suggested that this occurred via
the formation of the disulphonic acid 38, eqn. 13, and
concluded from other, mainly kinetic, evidence that the

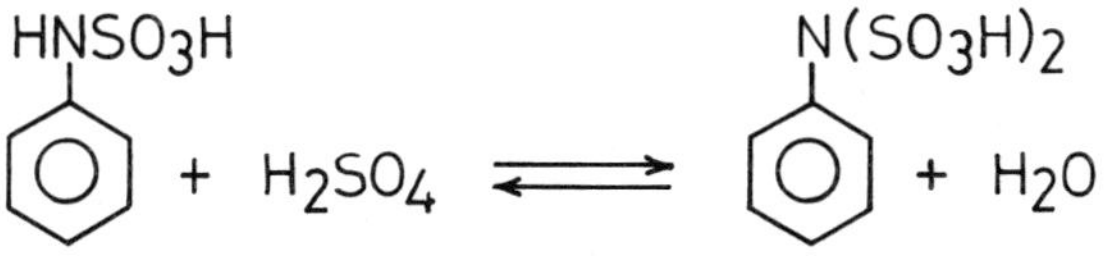

(13)

38

rearrangement is intramolecular with the intermediate formation of the ring-protonated bent boat structure, already
discussed as possibilities in the nitramine and nitrosamine
rearrangements. Studies in the naphthyl series, again using
[35]S as a label, have led Spillane, Scott and Goggin[56] to
propose that at least for 1-naphthylsulphamic acid (39) in
dioxan-sulphuric acid the rearrangement is partly intermolecular and partly intramolecular. Under these conditions
39 gave only 40; no trace was found of 41, the corresponding

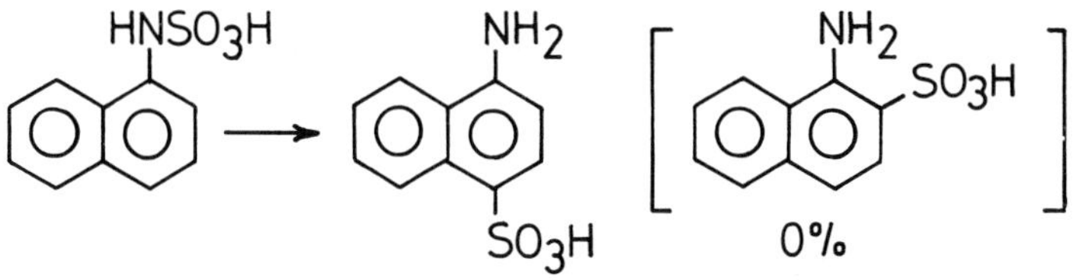

$$ (14) $$

ortho isomer. In the presence of [^{35}S]sulphuric acid the
recovered product 40 had only 62% of the activity expected
from a fully intermolecular process. A similar result was
obtained when the label was incorporated in the 1-naphthyl-
sulphamate. It was suggested that rapid desulphonation
followed by slow C-sulphonation gave rise to the fully
exchanged product, whereas the intramolecular pathway involves
the formation of some (unspecified) multiply sulphonated
derivative of 1-naphthylamine and involves a direct sulphonation by the sulphamate species rather than by sulphuric acid.
Strangely, Shilov and co-workers[57] had earlier reported that
the change 39→40 occurs at high temperatures, without pick up
of ^{35}S from the added labelled sodium sulphate. Clearly
further work is needed to clarify the mechanistic picture.

The related rearrangement of 2-naphthol-1-sulphonic acid
to give 2-naphthol-6-sulphonic acid has also been examined in
the presence of ^{35}S-labelled sulphonic acid, with interesting
results.[58] In 40-50% aqueous sulphuric acid, full uptake of
[35]S is observed; this is consistent with de-sulphonation
followed by re-sulphonation at carbon, an intermolecular

process. When, however, equimolar amounts of sulphuric acid
and substrate are used, there is no pick up of the label,
which suggests that now an intramolecular mechanism operates.
In glacial acetic acid containing sulphuric acid in excess, an
intermediate situation seems to exist, with both mechanisms
taking place. The results have been explained in terms of the
partitioning of the intermediate 42 (Scheme 11) to give the
final product by two separate routes.

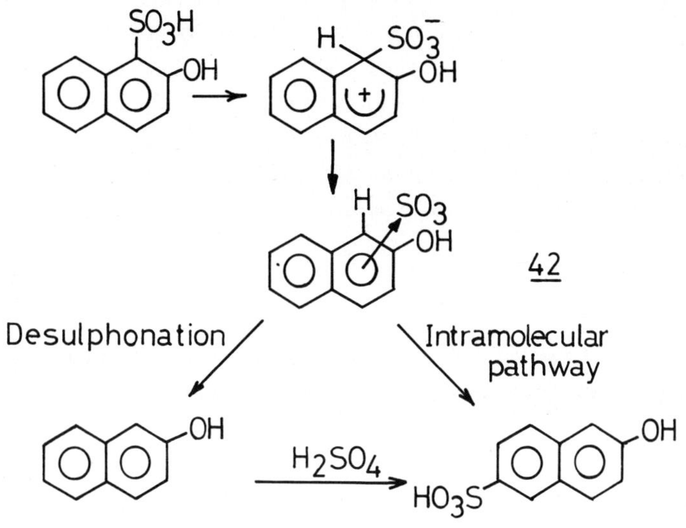

SCHEME 11

VII   REARRANGEMENT OF PHENYL ESTERS (FRIES REARRANGEMENT)

Phenolic esters when treated, in a variety of inert
solvents, or heterogeneously, with Lewis acids such as
aluminium chloride (or $SnCl_4$, $ZrCl_4$, $TiCl_4$, BF_3, $SbCl_5$, $ReCl_5$
etc.), rearrange to give ortho- and para- hydroxy substituted
phenyl ketones.[59]   Thus phenyl acetate (43) yields ortho- (44)
and para- (45) hydroxyacetophenone:

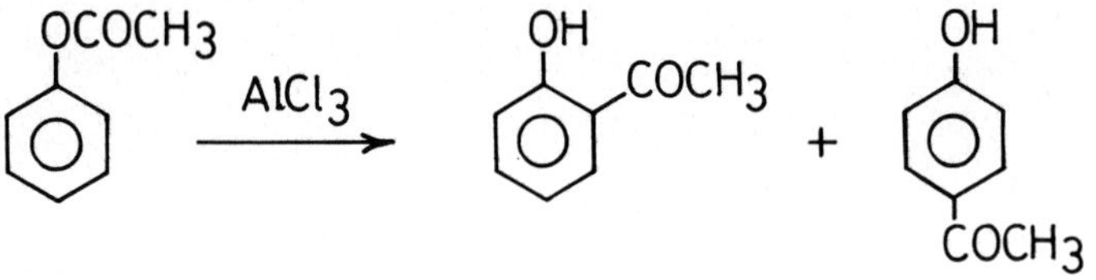

$$\text{OCOCH}_3 \quad \xrightarrow{\text{AlCl}_3} \quad \text{OH} \; \text{COCH}_3 \quad + \quad \text{OH} \; \text{COCH}_3 \tag{15}$$

43   44   45

The reaction has not been extensively studied from a mechanistic point of view, but it is a useful synthetic route to hydroxy ketones of this type. The ortho:para ratio of the products varies enormously with the reaction conditions such as temperature and solvent.[60] Generally the ortho product is favoured at higher temperatures. Measurement of the ortho:para ratio is complicated in some cases, particularly at higher temperatures, by the known conversion of the para product to the ortho isomer.[61]

There exists evidence in favour of all possibilities regarding the inter- or intra-molecularity of the change, viz that the reaction is entirely intramolecular,[62] entirely intermolecular[63] or concurrently both.[64] Most of this evidence is based on the ability or otherwise of trapping intermediates derived from a free acylating agent. One case for the intramolecular mechanism rests on the results of ^{14}C labelling experiments by Ogata and Tabuchi.[62] The reaction of phenyl acetate was carried out in the presence of [1-^{14}C]acetic anhydride. Acetate label was exchanged (eqn. 16) and both

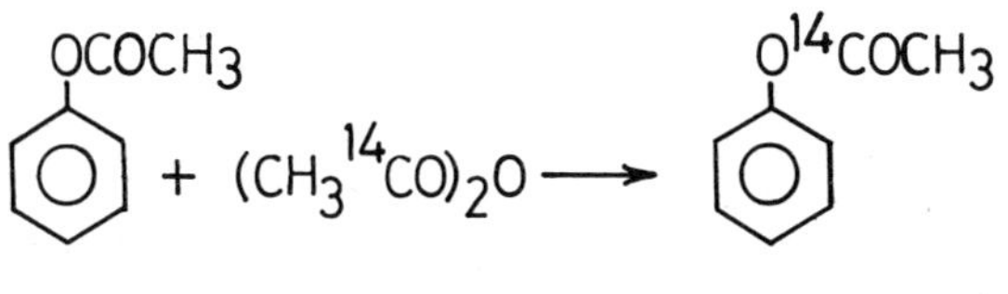

$$\text{OCOCH}_3 \quad + \quad (\text{CH}_3{}^{14}\text{CO})_2\text{O} \longrightarrow \text{O}^{14}\text{COCH}_3 \tag{16}$$

ortho-and para-hydroxyacetophenones recovered were radioactive. However, the extent of activity in the two products was the same, and significantly less than that in the recovered reactant. Similar results were obtained for three different solvent systems, 1,2 dichloroethane, nitrobenzene and light petroleum. The interpretation given was that intramolecular rearrangement occurs concurrently with acetate exchange in the reactant.

The proposed mechanisms for reaction intermolecularly all involve the formation of a free acylium ion, which then effects a conventional electrophilic substitution. Results obtained[64] from the variation of the rate constant with catalyst:ester ratio suggest that reaction can occur either via a 1:1 complex with the catalyst, or via a 2:1 complex. This has an analogy with the benzidine rearrangement where reaction paths via the mono- and di-protonated forms of the substrate have been identified (vide infra). It has been suggested[65] that the ortho product might arise from an intramolecular pathway and the para product from an intermolecular route as laid out in Scheme 12, although this has not been firmly established. It is probable that a range of mechanisms exist for this reaction, which operate for the different reaction conditions employed.

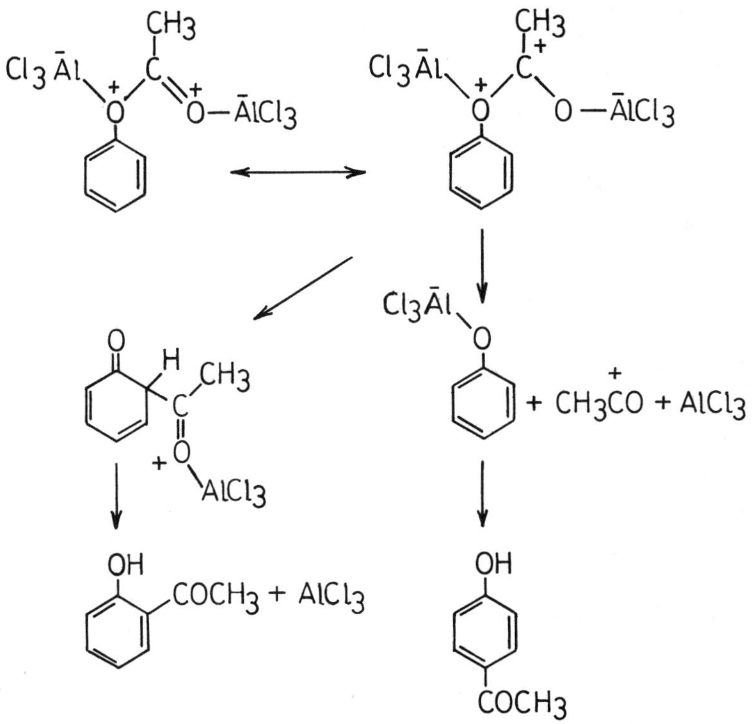

SCHEME 12

VIII   REARRANGEMENT OF ARYL ALKYL ETHERS

Aryl alkyl ethers (46) when treated either with strong acids or Lewis acids undergo a rearrangement, which is formally similar to the Fries rearrangement, to yield the

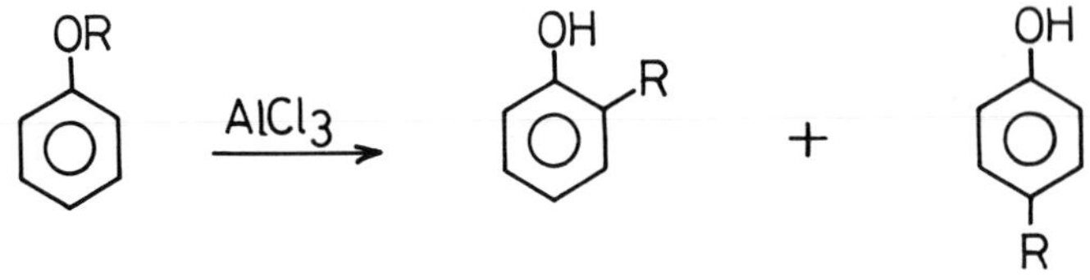

$$\text{(17)}$$

ortho- and para-alkylphenols 47 and 48, eqn. 17.   The phenol product of de-alkylation is often formed as a by-product. Relatively little mechanistic work has been carried out on this reaction, and again, as for the Fries rearrangement, the question of inter- vs intra-molecularity of the change is not wholly settled.   A high degree of retention of configuration (76%) has been found[66] when optically active para-tolyl l-phenylethyl ether is converted to ortho-phenylethyl-para-cresol.   This observation, together with results of earlier work by Wallis,[67] have been taken as strong evidence in favour of an intramolecular mechanism, at least for the formation of the ortho product.   Comparison of the product isomer ratios from the rearrangement with those from the Friedel-Crafts alkylation of phenols has been taken as support for this view.[68] Arguments have also been presented[69] for concurrent inter- and intra-molecular processes.

There is one report[70] of the use of an isotopic modification in a mechanistic investigation of this reaction, when the rearrangement of [2 - ^{2}H]s-butyl phenyl ether in chlorobenzene, using aluminium bromide catalyst, was examined. A mixture of ortho- and para-s-butylphenols and chloro-s-butylbenzenes were formed, in which deuterium scrambling had occurred between the 2- and 3- positions of the s-butyl group. This was interpreted in terms of a mechanism whereby fission of the butyl group occurred to form an ion pair, 49, which can

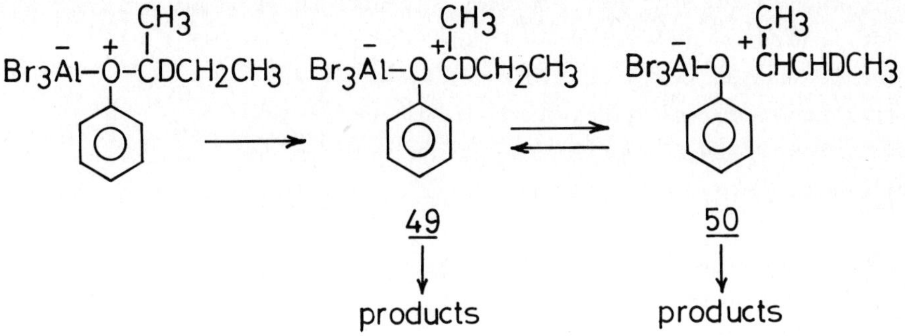

$$Br_3\bar{Al}\text{-}\overset{+}{\underset{|}{O}}\text{-}\overset{\overset{\displaystyle CH_3}{|}}{C}DCH_2CH_3 \quad Br_3\bar{Al}\text{-}\overset{+}{O}\overset{\overset{\displaystyle CH_3}{|}}{C}DCH_2CH_3 \quad Br_3\bar{Al}\text{-}O\overset{+}{}\overset{\overset{\displaystyle CH_3}{|}}{C}HCHDCH_3$$

49   50

products   products

SCHEME 13

undergo deuterium exchange within the ion pair via 50.
Earlier[71] an ion pair mechanism had been proposed from the
results of the reaction of optically active isobutyl phenyl
ether, when the reaction was carried out in the so-called
"inverse-addition" manner, ie. when a solution of the ether in
the reaction solvent was added to one containing the
equivalent amount of aluminium bromide in the same solvent.
When aluminium bromide is added to a solution of the ether in
the reaction solvent ("normal addition"), then an intermolecular mechanism is believed to predominate, in which a
molecule of the ether-$AlBr_3$ complex effects a direct
alkylation of a second ether molecule by an S_N2-type displacement.

When R = allyl group (in 46) the rearrangement is the
more familiar, and much more widely studied Claisen
rearrangement, yielding ortho- and para-allylphenols.
Considerable use of the ^{14}C isotope has been made in
mechanistic studies of this reaction, particularly under
thermal conditons. The elegant work of Schmid and co-workers[72]
has shown that under these conditions the reaction is fully
intramolecular, involving cyclic transition states, where the
terminal carbon atom of the allyl group in the reactant
becomes attached to the ortho-position of the aromatic ring in
the ortho-product. The para-isomer involves two such
cyclisations, so that the terminal carbon atom of the allyl

group in the reactant remains the terminal carbon atom of the allyl group in the para-allylphenol. Under these conditions, however, there is no evidence of any cationic intermediates and so the change falls outside the scope of this chapter. However, under quite different reaction conditions, i.e. in chlorobenzene solvent containing boron trichloride, and at quite low temperatures, rearrangement occurs in high yield, very readily, at rates estimated to be 10^{10} times greater than for the thermally induced reactions.[73] It is believed that the boron trichloride:ether complex(51) is first formed; rearrangement occurs by a [3s,3s] sigmatropic rearrangement, as for the thermally induced reaction, and the ortho-allyl product (52) is obtained after hydrolysis. The ^{14}C label in the γ-position in 51 is found quantitatively in the α-position

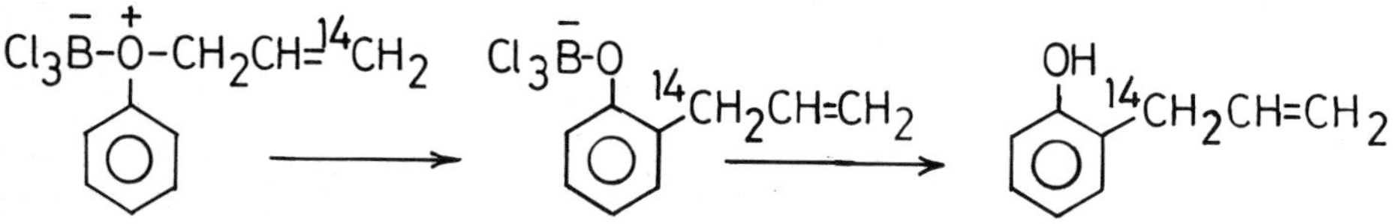

(18)

51   52

in 52. Experiments with branched allyl groups confirm that this is the predominant reaction, although small amounts of phenol products were detected, which could only have been formed by way of a [1s,2s] migration. Rearrangements other than the more usual [3s,3s] rearrangement have also been inferred from analysis of the position and extent of ^{14}C labelling in the products of the rearrangement of the di-ortho-substituted ether 53, terminally labelled. Apart from 54 (69% yield) which is formed by two [3s,3s] rearrangements, the meta substituted allylphenols were found, labelled in both α- and γ-positions. It is thought that 55 arises from successive [3s,3s] and [3s,4s] rearrangements and 56 from [3s,3s] and [1s,2s] changes. Schmid refers to the Claisen rearrangement brought about under these conditions as 'charge induced'.

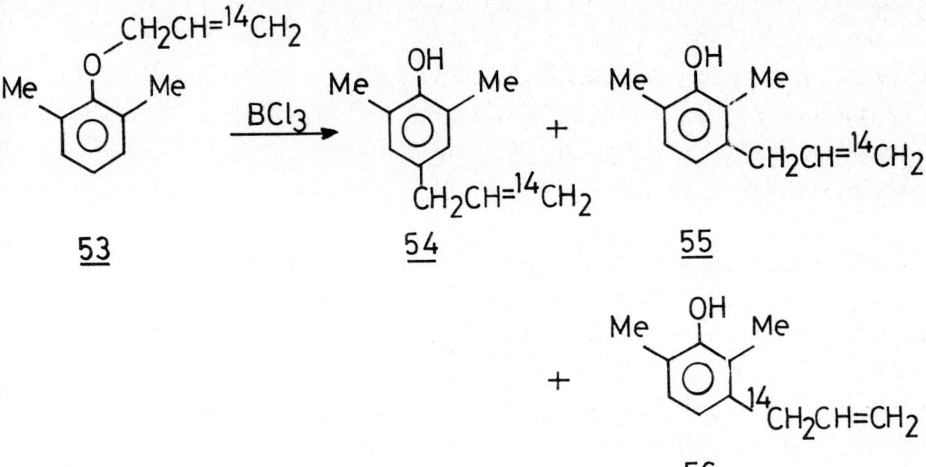

$$\text{(19)}$$

IX   REARRANGEMENT OF ALKYL AROMATIC HYDROCARBONS

Alkyl-substituted aromatic hydrocarbons when treated with acid catalysts (usually Lewis acids) often undergo rearrangement of the alkyl group to another position in the aromatic ring. An example of this[74] is the rearrangement of the tetramethylbenzenes prehnitene (57) and durene (58), both of which give isodurene (59) upon treatment with $HF-BF_3$, eqn. 20.

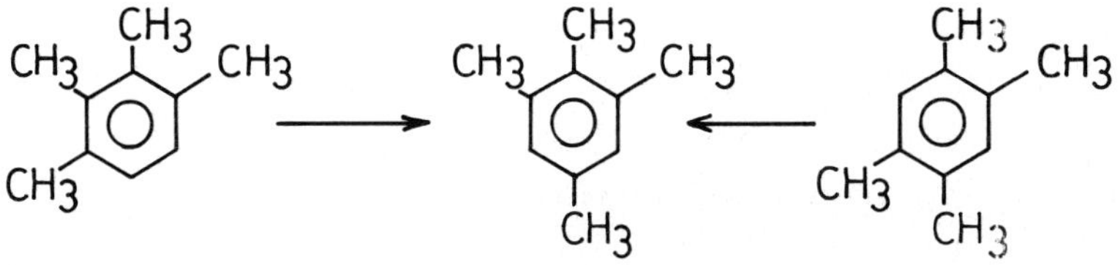

$$\text{(20)}$$

Sometimes alkyl migration accompanies some other reaction; e.g. the sulphonation of durene (58) gives prehnitenesulphonic acid (60),[75] eqn. 21. This reaction is an example of the Jacobsen rearrangement.

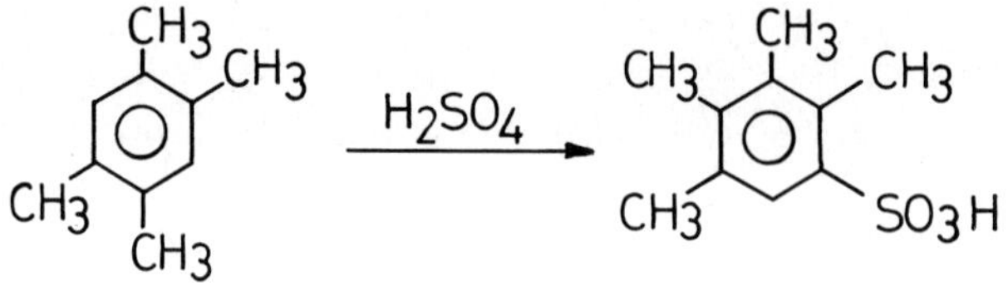

$$\text{(21)}$$

58   60

Methyl group migration in toluene has been examined[76] carefully using [1-^{14}C]toluene. When this is treated with Al_2Br_6-HBr at 20°C and 35°C, the activity of the 1-position in the aromatic ring decreases slowly as the methyl group becomes attached to other positions. At the same time, the activity of the 2-position increases to a maximum value and then decreases to a limiting value, whilst the activities of both the 3- and 4-positions gradually rise to their final values. These results strongly suggest a series of 1,2-methyl group rearrangements from 61 to 64, eqn. 22. In the naphthalene

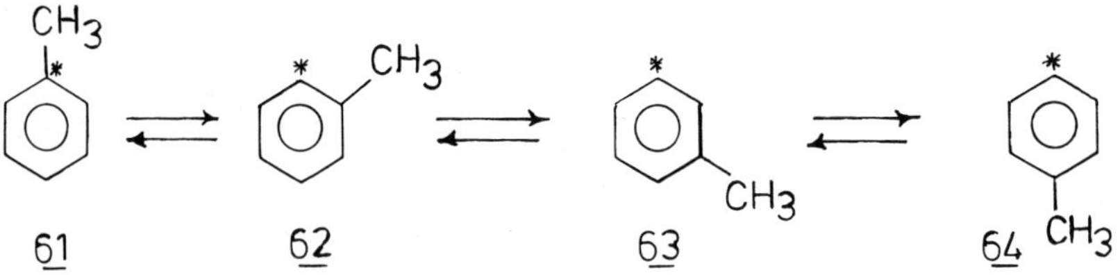

$$\text{(22)}$$

61   62   63   64

series, however, a similar experiment [77] using 1-methyl-[1-^{14}C]naphthalene shows that only 2-methyl[1-^{14}C]naphthalene is formed, i.e. only a single 1,2-methyl shift occurs. It is not clear why this situation is different from the toluene case.

Whilst the use of ^{14}C ring-labelled materials clearly shows that these 1,2 migrations occur, the mechanism of such changes is not unambiguously established. It is generally believed that reactions go via the σ-complex 65, even though n.m.r. studies have shown[78] that 66 and 67 are the predominant species obtained upon protonation of the aromatic system in high acid media. It is claimed by some workers that 65 is converted to the π-complex 68 during the rearrangement,

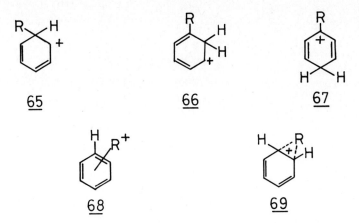

65   66   67

68   69

whereas others invoke the intermediacy of a bridged ion such
as 69, whilst yet another view is that there is no intermediate
at all, and structure 69 represents a situation close to the
transition state for the 1,2 migration. (See ref. 65 p. 14).

Generally in alkylbenzenes other than toluene, a further
complication arises in that disproportionation can readily
occur in many cases alongside an intramolecular rearrangement.[79]
Rearrangement also is not confined to intramolecular 1,2
migrations but also occurs by direct transfer of the alkyl group
from one molecule to another, as in the case of direct ortho
to para transfer in xylenes, ethyltoluenes and cymenes.[79]
This has clearly been demonstrated in the case of the reactions
of ethylbenzene[80] in the presence of $AlBr_3$-HBr, when benzene,
diethylbenzene and triethylbenzene are the products.
Experiments with ethyl[1-^{14}C]benzene have thrown much light
on the role of disproportionation and direct ethyl group
transfer in alkyl group rearrangements of this type. It is
worth noting at this point that large errors can arise in the
calculation of the extent of labelling in the various
positions in the aromatic system, depending on the method of
degradation and analysis used. An improved degradation
procedure and calculation[81] on the ethylbenzene system
reduced these uncertainties drastically. It was established
that the ion derived from para-diethylbenzene is important
in this reaction as an alkylating agent itself. The following
reactions were identified in this connection.

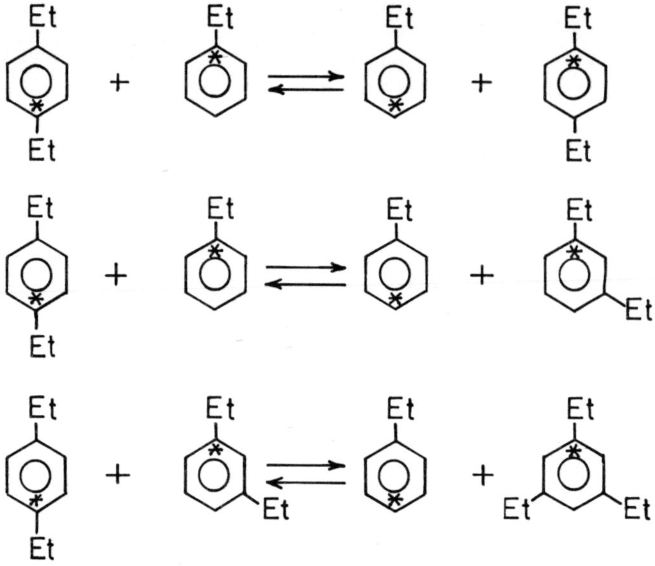

SCHEME 14

It had been suggested earlier[82] that the transition state of such a bimolecular reaction of a σ-complex with another aromatic molecule would resemble 70. An alternative view

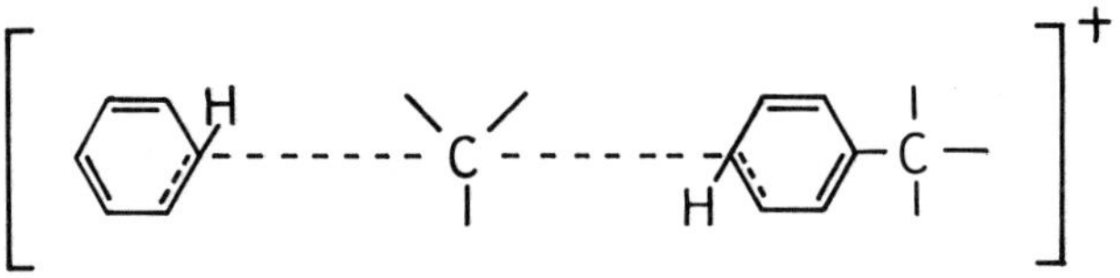

70

of disproportionation and direct alkylation has been presented by Streitwieser and Reif[83] who consider that an alkyl carbonium ion is involved. This conclusion derives from the results of a double-labelling experiment. When optically active $(\alpha-^2H_1)$-ethylbenzene, ring labelled with ^{14}C (71), was treated with $GaBr_3$-HBr in benzene solvent at 50°, the ethyl group became transferred to the solvent, since the radioactivity level of ethylbenzene was gradually reduced with time. Furthermore, the optical activity was also lost and deuterium exchange occurred with the formation of the non-deuterated ethylbenzene

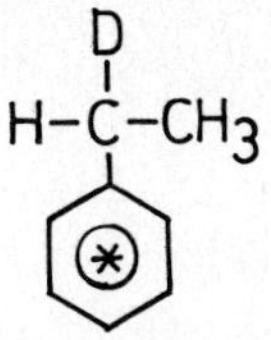

<u>71</u> 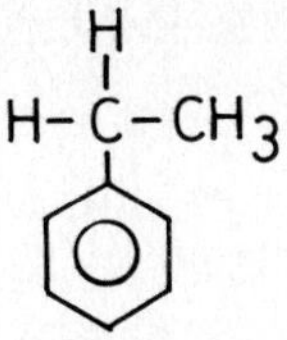   <u>72</u> 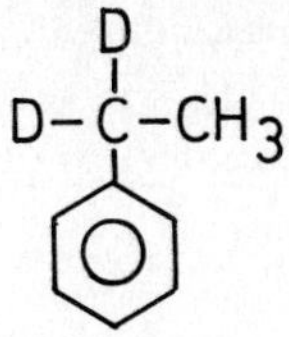   <u>73</u>

<u>72</u> and also the di-deuterated form <u>73</u>. These observations
cannot be accounted for by S_N2 substitution in a σ-complex,
but can readily be interpreted in terms of a hydride
abstraction reaction to form the carbonium ion <u>74</u>, thus
bringing about hydrogen exchange. This is followed by an
alkylation of the solvent, an acid-promoted de-alkylation of
the resulting 1,1-diphenylethane and a final hydride
abstraction. The sequence is set out in Scheme 15.

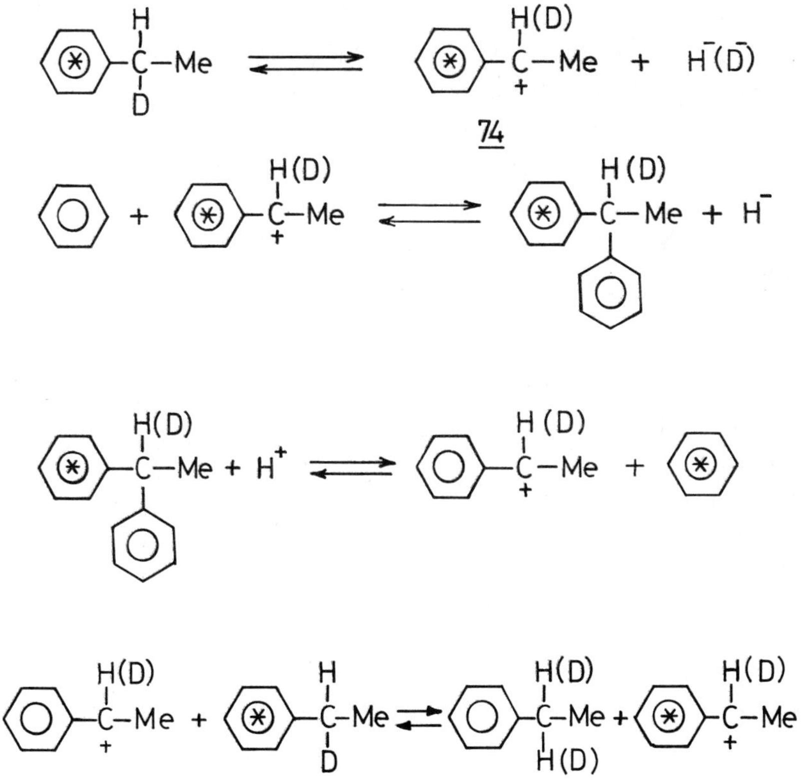

SCHEME 15

X THE WALLACH AND RELATED AZOXY REARRANGEMENTS

A. Structural Aspects

The rearrangement of azoxybenzene 75 to p-hydroxyazo-benzene 76 which occurs in concentrated sulphuric acid media is known as the Wallach rearrangement:[84]

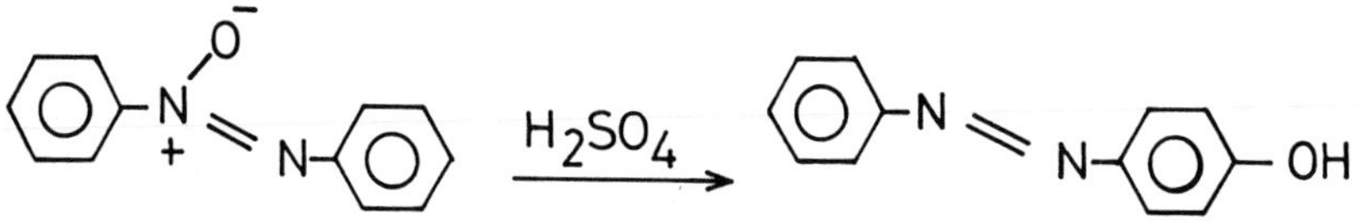

(23)

75   76

Related processes which are currently considered as falling under this heading include reactions in chloro- or fluorosul-phonic acid, yielding the corresponding azoaryl halosulphates, and with sulphonic acid anhydrides and chlorides, yielding the sulphonate esters. Ortho - hydroxyazo products are obtained to extents varying from very small (or zero) with azoxybenzene itself, to predominant in the case of p,p'-disubstituted derivatives. A photochemical rearrangement is also known, giving rise exclusively to o-hydroxyazobenzene, but this process will not be considered further in this chapter (see however refs. 85,86).

Systematic studies with substituted azoxybenzenes have been relatively few in number, especially so with unsymmetrica-lly substituted derivatives. A pertinent finding with the latter series is that isomeric α- and β-azoxybenzenes* (77, 78)

---

*An alternative nomenclature uses the following numbering system for azoxybenzene.

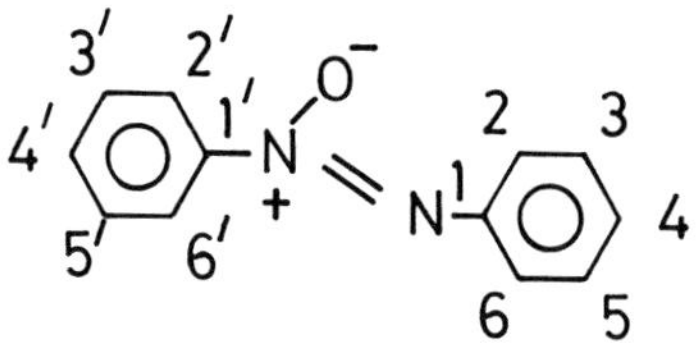

generally yield the same product on rearrangement, with OH
entering the unsubstituted phenyl ring, e.g. eqn. 24,[87]
although some exceptions to this have been noted.[88] It should
be mentioned that azoarenes are also obtained in some
reactions,[89] occasionally as the major products (e.g. with
4,4'-dibromo- or diiodoazoxybenzene,[90] but there has been no
detailed study of such processes. With azoxybenzene itself, in
the dilute solutions used in kinetic studies (vide infra),

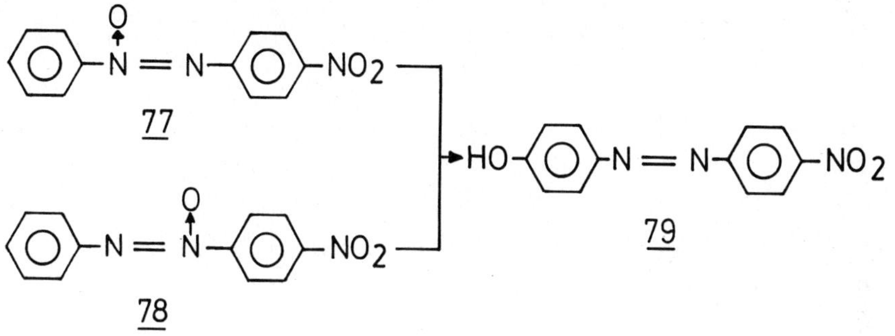

$$\text{(24)}$$

there is negligible formation of azobenzene.

B. Results of isotopic tracer studies with azoxyarenes

Mechanistic investigations of the Wallach rearrangement
began with isotopic tracer studies; these were directed
initially at the question of detection of possible reaction
intermediates. The first studies, carried out by Shemyakin
and co-workers,[91] involved the rearrangement of azoxybenzene
specifically labelled on one nitrogen with ^{15}N (80). The fate
of the label in the product was determined by reductive
cleavage of the azo linkage followed by acetylation and
isotopic analysis of the separated acetanilides (Kjeldahl
degradation to NH_3 and conversion to N_2). The result of the
experiment was that the two aniline residues produced on
rearrangement each contained close to half the isotopic enrichment of the reactant, i.e. equalization of the label had
essentially taken place:

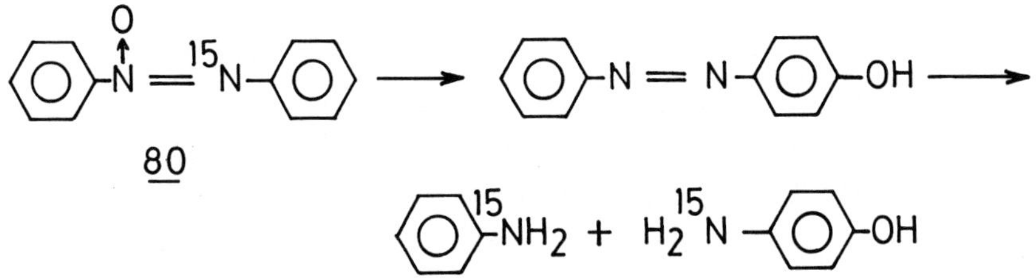

$$\text{(25)}$$

Importantly, the recovered unreacted azoxybenzene contained the label undisturbed. It was concluded[91] that the reaction pathway involves formation of a symmetrical intermediate, the N,N-oxide species (81), which would be attacked on either ring with equal ease:

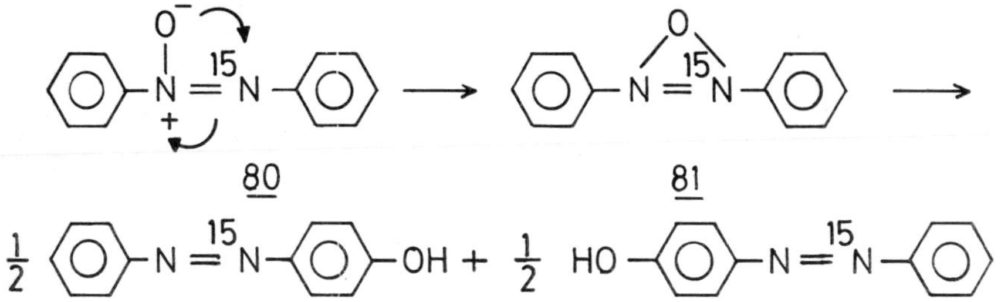

$$\frac{1}{2}\ \text{(O)}-N \overset{15}{=} N-\text{(O)}-OH + \frac{1}{2}\ HO-\text{(O)}-N \overset{15}{=} N-\text{(O)} \tag{26}$$

The fact that the recovered reactant contained the label undisturbed indicated that the symmetrical intermediate was not formed in a pre-equilibrium prior to the rate controlling step.

A similar tracer experiment was performed subsequently by Behr and Hendley[92] using azoxybenzene specifically labelled with ^{14}C at the 1-position (82). This study likewise showed that the label was practically equalized in the product:

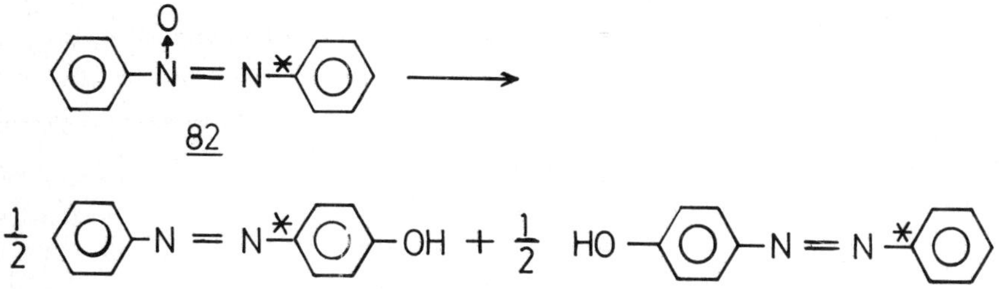

$$\frac{1}{2}\ \text{(O)}-N = N\overset{*}{-}\text{(O)}-OH + \frac{1}{2}\ HO-\text{(O)}-N = N\overset{*}{-}\text{(O)} \tag{27}$$

However, there was a slight preponderance in favour of one of the species, as it was found that azoxybenzene having an activity of 0.3550 μ Curies/mmole on treatment with 83% H_2SO_4 at 90° for 30 min yielded aniline with 0.1730 μ Curies/mmole and p-aminophenol with 0.1627 μ Curies/mmole. The ^{14}C tracer study also eliminated the possibility that during the course of reaction the nitrogen atoms might become separated from the aromatic rings, which could not have been discerned in the ^{15}N tracer study. The finding of rearrangement in diazonium ions (vide infra) brings such a pathway within the realm of possibility.

Whether the Wallach rearrangement occurs by intra- or inter- molecular pathways has been answered by means of ^{18}O tracer studies[93-95] using azoxybenzene labelled with ^{18}O, or isotopically normal azoxybenzene and sulphuric acid enriched with ^{18}O. In either case, it has been found that the oxygen in the p-hydroxyazobenzene product is derived from the solvent, while the oxygen in the o-hydroxyazobenzene (formed as minor product) contains the isotopic composition of the parent azoxybenzene.* However, with some of the substituted azoxy-benzenes the situation is not clear-cut. Thus with 4,4'-dimethylazoxybenzene the ortho-hydroxyazo product has the label derived in part from the solvent and in part from the parent substrate (eqn. 28), indicating that two reaction pathways co-exist:[94]

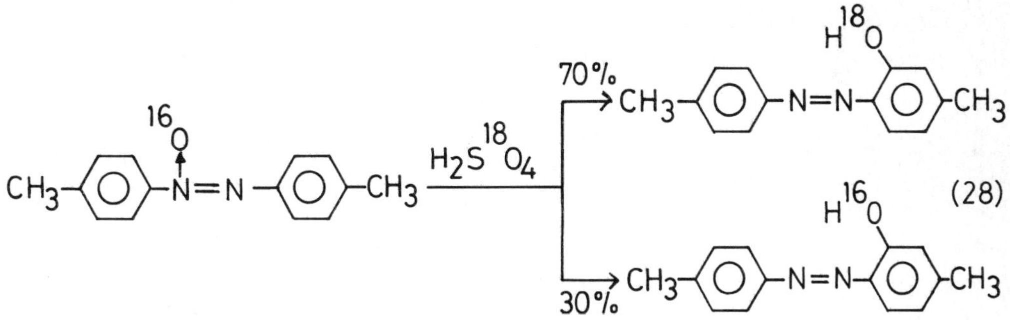

(28)

Intramolecularity in the ortho rearrangement has been accounted for by cyclic process (eqn. 29),[94] though Oae et al.[93,96] favour an intimate ion pair species (84) in place of the cyclic intermediate (83).

---

* Certain hydroxyazo compounds have been found to undergo exchange of the hydroxyl oxygen in acidic medium.[96]

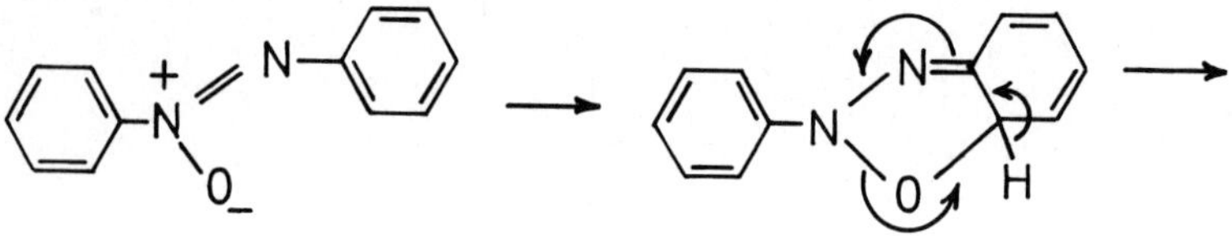

$$76 \quad (29)$$

83

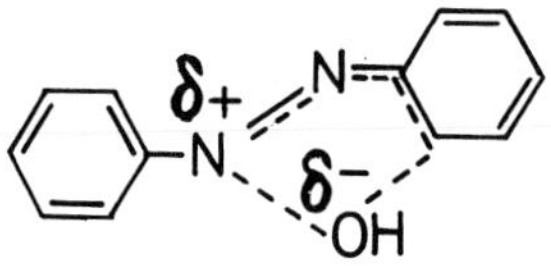

84

C. The β→α isomerisation

The asymmetry of the azoxy function, which gives rise to
structural isomerism of azoxyarenes (eqn. 24), raises the
possibility of interconversion among such isomers. Only one
case has been found so far where this isomerisation occurs
under the conditions of the Wallach rearrangement, i.e. that
of β-p-nitroazoxybenzene (78) which isomerises to the α-form
(77) in 98% H_2SO_4.[97] Reaction of β-p-nitroazoxybenzene in
$H_2S^{18}O_4$ showed that the p-hydroxy-p'-nitroazobenzene product
derives its oxygen from the medium, as expected. However, the
azoxy compound recovered from the reaction mixture following
partial rearrangement was the isomerised α-form (77) whose
oxygen was isotopically normal.[94] It was tempting to assume
that the β→α isomerisation and the Wallach rearrangement occur
by a common intermediate, the N,N-oxide.[95]

An alternative mechanism for the β→α isomerisation may be
proposed, however, as shown in Scheme 17.[98] The starting point
in this scheme is the substrate protonated on the NO_2 function,
in accord with the observation that isomerisation requires
highly concentrated acid media. The bridged structure is
expected to be formed as a short lived intermediate, or merely
as a transition state.

It would be desirable to perform a kinetic study of the
concurrent processes, the β→α isomerisation and the Wallach
rearrangement, noting that kinetic studies of the latter have
been quite informative.

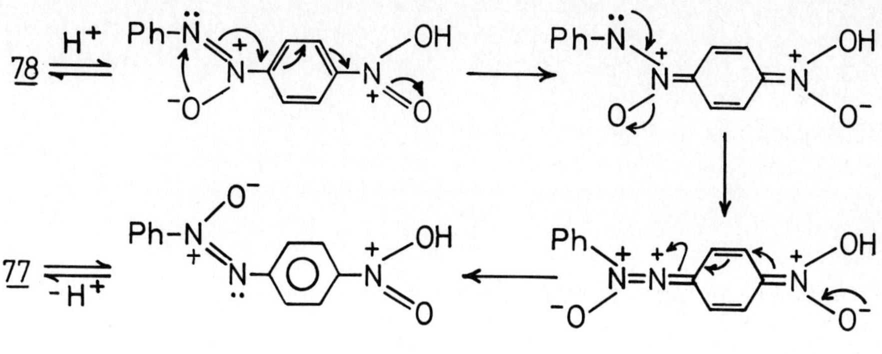

SCHEME 17

D. Results of kinetic studies with azoxyarenes

The isotopic tracer studies of the Wallach rearrangement, while highly informative regarding the question of symmetrical intermediates and the inter- or intramolecularity of the over- all process, provided little insight into the role of the acid medium or the reason for the requirement of concentrated acid media, rather than dilute acid media which suffice in, say, the benzidine rearrangement (vide infra). A kinetic study was thus undertaken in our laboratory[99] and has provided information complementary to that derived from the isotopic tracer studies.

The results of a spectrophotometric study of the trans- formation of azoxybenzene to p-hydroxyazobenzene in aqueous sulphuric acid media are given in Table 5.[100] Included in this table are equilibrium protonation data for azoxybenzene as evaluated from the measured pK_a of the substrate. Also given are the Hammett acidity function values for these acid media. It is immediately apparent that from the data that azoxybenzene is largely protonated already in 75% H_2SO_4, but the rate of rearrangement continues to increase beyond the stage of complete monoprotonation of the substrate. This behaviour points to the requirement of a second proton transfer.

To accommodate the requirement of a two-proton process, a reaction scheme may be proposed as follows (Scheme 18). Initially there is a rapid equilibrium protonation of azoxybenzene to yield the conjugate acid 85, which is the effective reactant. This is followed by one of the routes

TABLE 5

Kinetic data for rearrangement of azoxybenzene to 4-hydroxyazobenzene in aqueous H_2SO_4 at 25°.[101]

H_2SO_4 weight %	$-H_o{}^a$	$\dfrac{c_{SH^+}{}^b}{c_S + c_{SH^+}}$	$10^5\ k_\psi{}^c$ s^{-1}
75.30	6.65	0.967	0.016
80.15	7.42	0.994	0.208
85.61	8.35	0.999	2.17
90.37	9.05	1.000	7.23
95.19	9.82	1.000	20.9
97.78	10.35	1.000	43.8
99.00	10.82	1.000	76.8
99.59	11.18	1.000	227
99.90	11.64	1.000	860
99.97	11.84	1.000	2310
99.99	11.90	1.000	4160

[a] Data from ref. 146.

[b] Calculated using pKa for azoxybenzene -5.15

[c] Pseudo first order rate constants as determined spectrophotometrically

indicated in Scheme 18, both pathways leading to the dicationic intermediate 88. Attack by nucleophile on the para position of 88, followed by proton loss, yields the desired product. It is noted that if instead of H_2O one considers HSO_4^- as the nucleophilic species, then the initially formed azoaryl hydrogen sulphate would be expected to be rapidly hydrolysed to the azophenol 76.[101] The latter will also exist in the protonated form in these acidic media.[102]

Of the two possible pathways indicated in Scheme 18,
i.e. formation of a discrete diprotonated species 86, followed
by loss of H_2O as the slow step, _versus_ rate determining
proton transfer occurring synchronously with heterolysis of the
N-O bond through participation of the nitrogen lone pair of the
transition state 87, we currently favour the latter pathway.

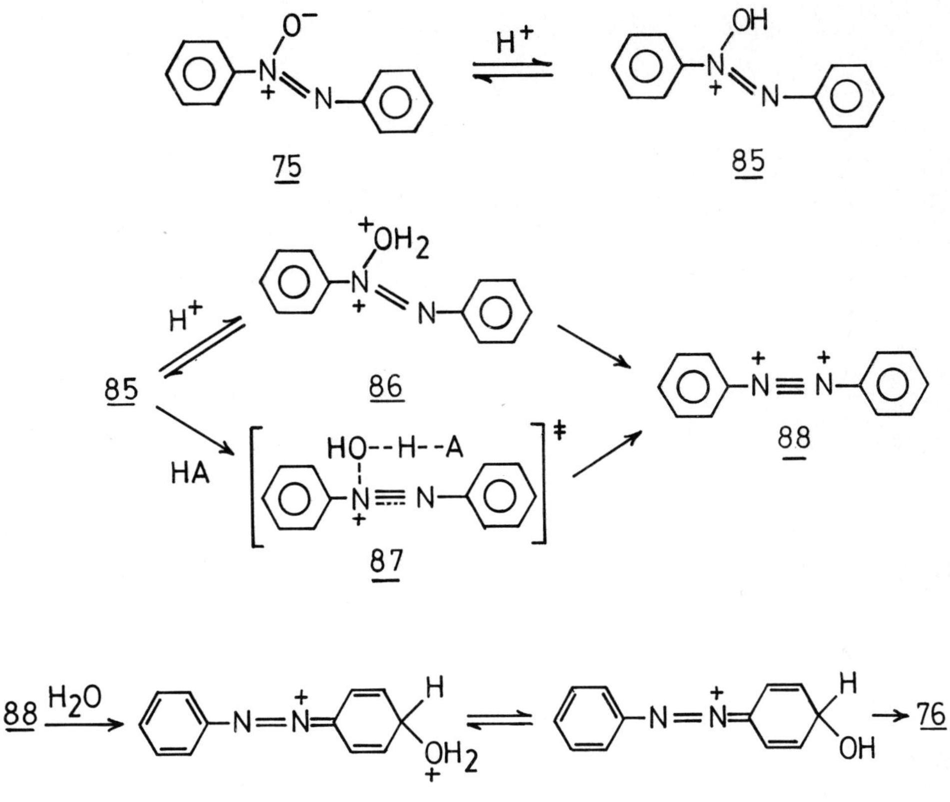

SCHEME 18

This is based on the kinetic analysis given in eqns. 30-34 ($S = \underline{75}$, $SH^+ = \underline{85}$, $X^{2+} = \underline{88}$, which predicts linearity in the plot of $\log k_\psi - \log (C_{SH}+)/(C_S + C_{SH}+)$ vs. $\log a_{HA}$, as actually found in Fig. 1. The observed linear correlations are thus in accord with general acid catalysis by undissociated H_2SO_4 and $H_3SO_4^+$ species.[104]

$$S + H^+ \rightleftharpoons SH^+ \quad \text{(pre-equilibrium)} \tag{30}$$

$$SH^+ + HA \longrightarrow X^{2+} \quad \text{(rate-limiting)} \tag{31}$$

$$X^{2+} - - \rightarrow \text{product (fast steps)} \tag{32}$$

$$\text{rate} = -\frac{d(C_S + C_{SH}+)}{dt} = k_\psi(C_S + C_{SH}+) = k'_o \frac{a_{SH}+ \, a_{HA}}{f_{\neq}} \tag{33}$$

$$\log k_\psi - \log \frac{C_{SH}+}{C_S + C_{SH}+} = \log a_{HA} + \log k' + \log \frac{f_{SH}+}{f_{\neq}} \tag{34}$$

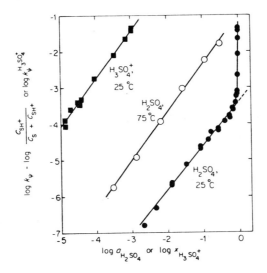

Figure 1. Graphs of $\log \underline{k}_\psi - \log (\underline{C}_{SH}+/\underline{C}_S + \underline{C}_{SH}+)$ at $25^\circ C$ ($\bullet$) and $75^\circ C$ (O) vs. $\log \underline{a}_{H_2SO_4}$, and of $\log \underline{k}_\psi$ at $25^\circ C$ ($\blacksquare$) vs. $\log \underline{X}_{H_3SO_4}+$, for the rearrangement of azoxybenzene in H_2SO_4.[101]

The dicationic intermediate $\underline{88}$ [97] would clearly account for equalization of the ^{15}N and ^{14}C labels in the above

described tracer experiments. The preference for para attack
on 88 by H_2O (or HSO_4^-) is in accord with calculations of
charge densities.[105]   The dicationic species 88, as well as
the corresponding 4,4'-dichloro analog, have been observed by
n.m.r. spectroscopy in the SbF_5-HF-SO_2 system,[106] though this
does not prove that such species are present as reaction intermediates under typical Wallach rearrangement conditions.
Other symmetrical species that have been advocated as reaction
intermediates in the Wallach rearrangement are the mono- and
di-O-protonated N,N-oxide species 89 and 90, and the di-N-
protonated species 91.[93,95,107]

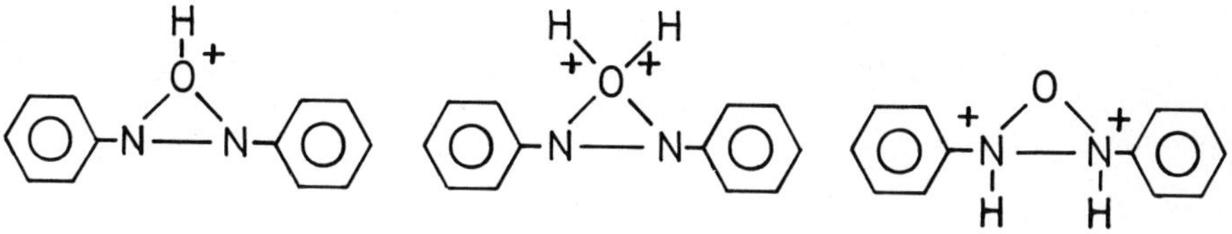

89   90   91

     Reaction pathways involving unsymmetrical species as the
key intermediates in Wallach transformations have also been
postulated.   In the case of the azoxybenzene rearrangement the
quinonoid species 92 and 93 have been advanced:[108]

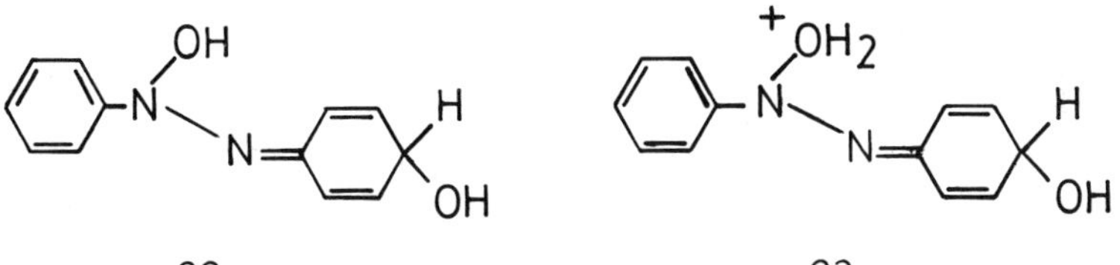

92   93

The results of the ^{15}N and ^{14}C experiments would then be
accommodated by postulating that nucleophilic attack would be
equally facile on either of the phenyl groups.   In basic medium,
4- and 4'-bromoazoxybenzene have in fact been found to liberate
bromide ion at about equal rates.[109]   The lack of a kinetic
isotope effect in the rearrangement of azoxybenzene-d_{10} in
H_2SO_4[109] shows that removal of the aromatic hydrogen from 92
and 93 is not rate-limiting , but this is implicit also in the

mechanim shown in Scheme 18.

Reaction pathways involving both dicationic and quinonoid intermediates have been postulated for rearrangement in the naphthyl azoxy series[110] and in the case of 2,2',4,4',6,6'-hexamethylazoxybenzene (eqn. 35).[111] However, isotopic studies have not been performed on these systems so far.

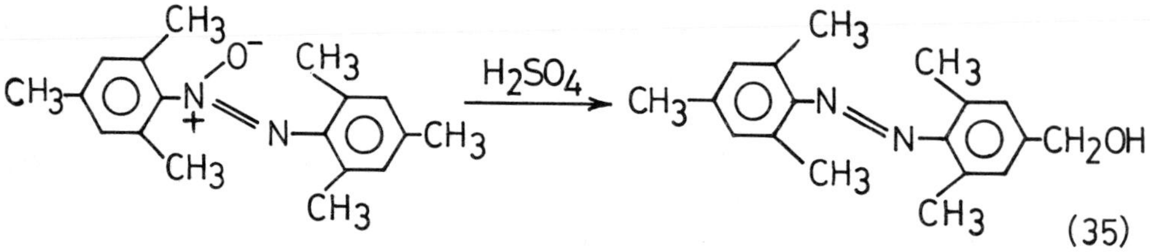

(35)

E. Reactions of azoxyarenes with arenesulphonic anhydrides and chlorides

Reaction of azoxyarenes with arylsulphonic anhydrides in benzene or acetonitrile as solvent results in Wallach type rearrangement to give the p-arenesulphonyloxyazobenzene and arenesulphonic acid, e.g. eqn. 36.[112]

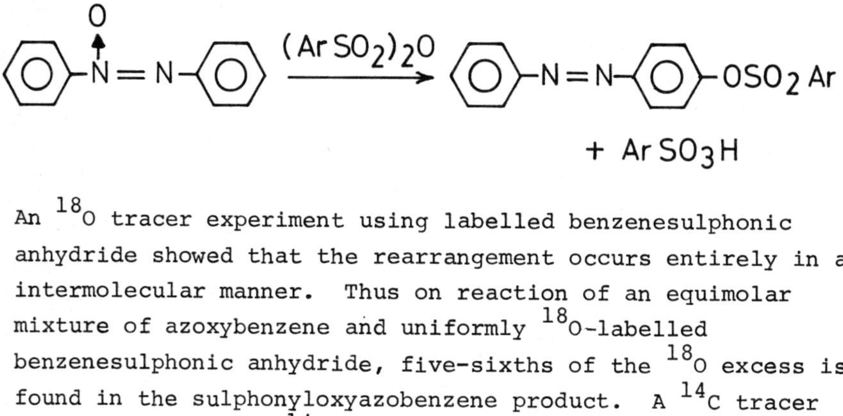

(36)

$$+ \ Ar\, SO_3H$$

An ^{18}O tracer experiment using labelled benzenesulphonic anhydride showed that the rearrangement occurs entirely in an intermolecular manner. Thus on reaction of an equimolar mixture of azoxybenzene and uniformly ^{18}O-labelled benzenesulphonic anhydride, five-sixths of the ^{18}O excess is found in the sulphonyloxyazobenzene product. A ^{14}C tracer study using azoxy[1-^{14}C]benzene revealed that attack of the arenesulphonate on the ring removed from the N(O) function was slightly favoured compared to attack on the ring adjacent to the N(O) function. It was concluded that the main reaction pathway involves formation of a symmetrical intermediate, for

which the sulphonylated N,N-oxide species $\underline{94}$ was proposed
(Scheme 19); the minor competing pathway was not identified,
however.

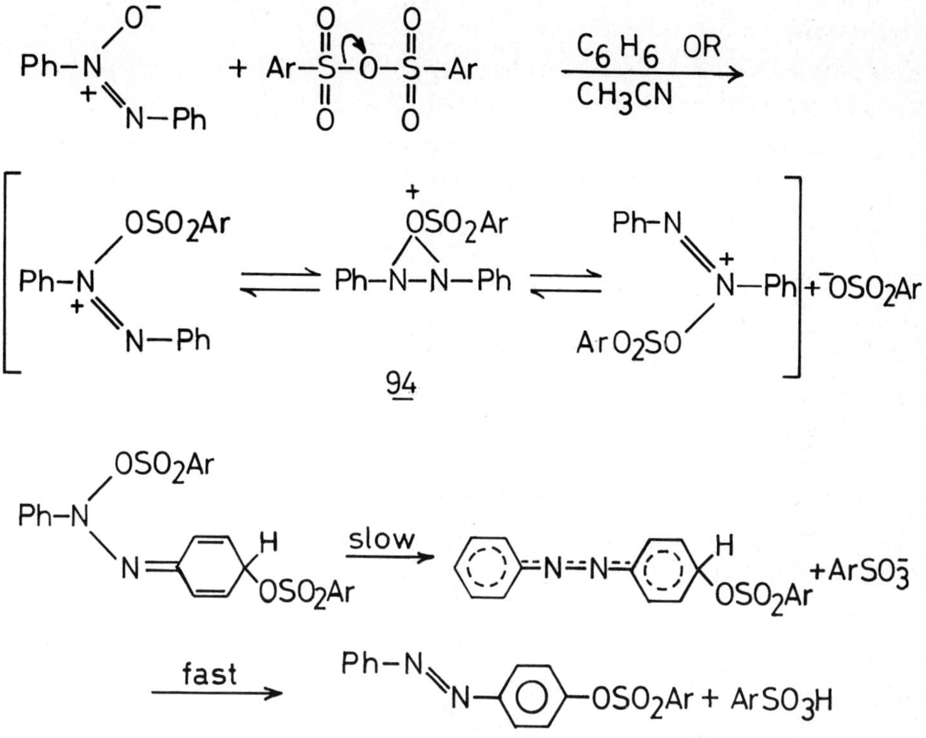

SCHEME 19

Reaction of azoxybenzene-$\underline{d}_{10}$ with benzenesulphonic
anhydride showed no kinetic isotope effect ($k_H/k_D = 1.00 \pm$
0.05 in CH_3CN at 60°), indicating that aryl proton loss is not
rate determining, in accord with the above scheme. Kinetic
study of benzenesulphonic anhydride with substituted
azoxyarenes (CH_3CN, 60°) yielded a linear correlation between
log k and the Hammett σ substituent values. Since in the
case of OMe the σ^+ value was not required, it was concluded
that a dicationic intermediate $\underline{88}$ was not involved in these
processes.[112]

Azoxybenzene reacts with arenesulphonyl chlorides to yield
ortho- and para-arenesulphonyloxyazobenzenes (eqn. 37); the

ratio of products could be accurately determined by means of isotopic dilution techniques on using azoxy[1-^{14}C]benzene as the reactant.[113]

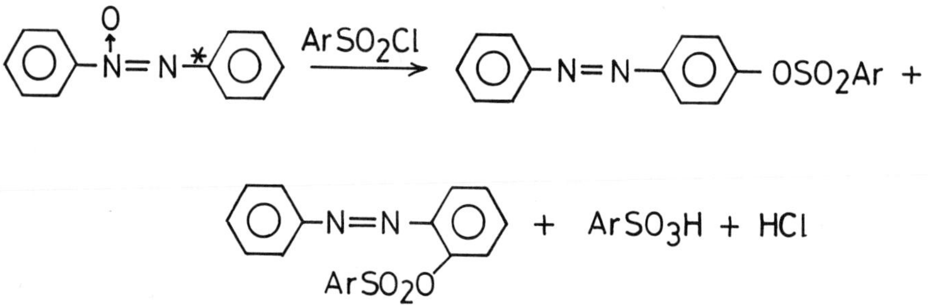

$$\text{(37)}$$

The tracer results also showed that the sulphonyloxy group migrates to both phenyl rings. Moreover, the migration proceeds ca. two-thirds to the ortho position of the ring removed from the N(O) function and one-third to the ortho position of the ring adjacent to the N(O) function. In other tracer experiments using ^{18}O-labelled arenesulphonyl chlorides, it was found that the ortho rearrangement occurs intramolecularly and the para rearrangement intermolecularly. The reaction mechanism proposed for these processes is shown in Scheme 20.[113] Noteworthy is the bridged species 95 which is proposed to account for the intermolecular nature of the ortho migration. On the other hand, formation of the ion pair species 96 followed by nucleophilic attack would account for the intermolecularity in the para rearrangement.[113]

It is apparent that the arenesulphonic anhydride/ chloride azoxyarene reactions have extended the Wallach rearrangement process in a most useful manner. One can look forward to future studies to unravel further the competing pathways indicated by the isotopic tracer results.

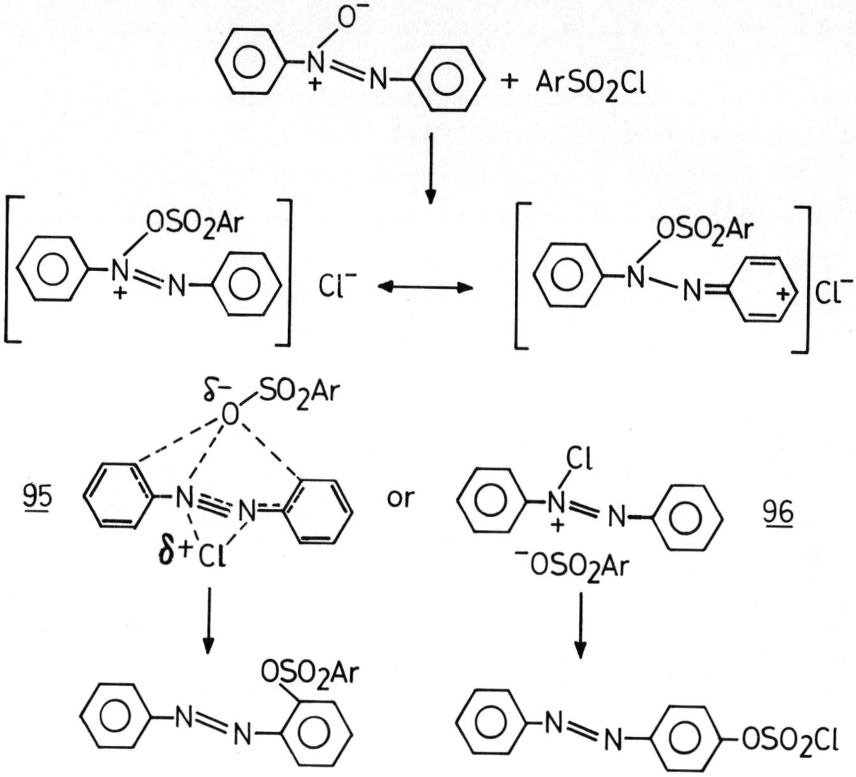

SCHEME 20

XI BENZIDINE REARRANGEMENT

A. General aspects

The conversion of hydrazobenzene (97) to benzidine (98) in solutions of organic solvents containing mineral acid was first reported in 1863,[114] while some 15 years later the formation of diphenyline (99) was noted as a minor product, eqn. 38:[115]

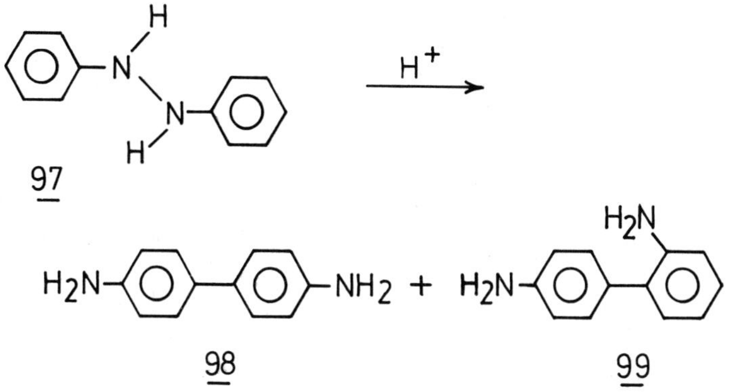

(38)

Extensive product analyses have since revealed that, under
appropriate reaction conditions, e.g. action of HCl gas on
solid 97,[116] still other isomeric products are formed in
lesser proportions, namely, ortho-benzidine (100) and para-
and ortho-semidines, (101) and (102).[117]

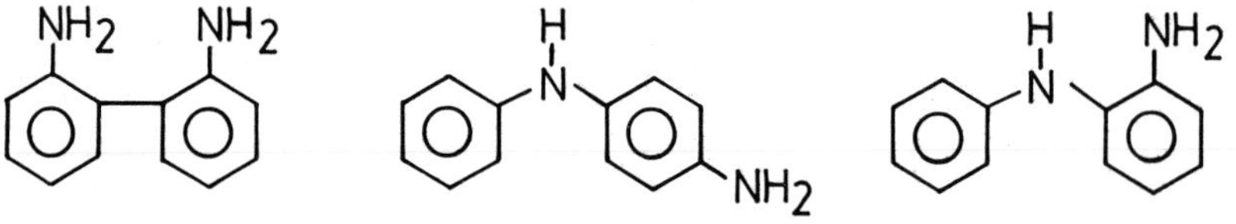

100   101   102

In addition, disproportionation can occur to varying extents
according to eqn. (39):

$$2 \ ArNHNHAr \xrightarrow{H^+} ArN=NAr + 2ArNH_2 \qquad (39)$$

All the above processes comprise the acid catalysed benzidine
rearrangement. Thermal and photochemical processes are also
known[118,119] but these fall outside the present treatment
which is limited to cationic processes.

The effect of structural changes has been widely
investigated, as for example in the naphthyl series, or in
the phenyl series via substituents on the phenyl groups as
well as on one or both nitrogens, including bridged structures.
Kinetic and isotopic studies have featured importantly. All
this derived information has not, so far, yielded a unique
picture of the reaction; on the contrary, mechanistic
proposals have proliferated to the extent that there are at
least four major theories that are current, while many
proposals have become outdated by new evidence. In 1962,[120]
Ingold was led to indicate the transition state of the
benzidine rearrangement as being in the clouds. While
important questions have been answered since that time, others
have been raised, and a real solution to the problem still
appears distant. Though this situation is a challenging one
to the researcher, the task of the reviewer is indeed
difficult, especially when faced with space restrictions.

B.   The one- and two-proton processes

The dominant finding bearing on the mechanism of the benzidine rearrangement against which all other criteria must be measured derives from kinetic observations pointing to the existence of one- and two-proton parallel pathways[121-124] (cf. the Wallach rearrangement). Thus the general rate law applicable to the rearrangement of hydrazoarenes (S) is given by eqn. 40 which can be re-written as eqns. 41 and 42.

$$\frac{-d[S]}{dt} = k_2[S][H^+] + k_3[S][H^+]^2 \qquad (40)$$

$$\frac{-d[S]}{dt} = k_T[S] \qquad (41)$$

$$k_T = k_2[H^+] + k_3[H^+]^2 \qquad (42)$$

It is seen that a plot of log k vs. log $[H^+]$ (or $-H_o$) will be a curve with an initial slope close to 1 and a final slope close to 2, while a plot of $k_T/[H^+]$ vs. $[H^+]$ will be linear with intercept k_2 and slope k_3.

Thus, depending on its structure, a given hydrazoarene may react by a rate law which is first order in $[H^+]$ over the total range of acidity, second order in $[H^+]$, or simultaneously by both processes. Apparent non-integral orders in $[H^+]$ can then result when measurements are performed over a relatively narrow acid concentration range.

In Table 6[125-137] are given the pertinent results for some selected compounds derived according to the above kinetic analysis. In the case of hydrazobenzene and para-hydrazotoluene it was shown that the same rate law governed the formation of all products (benzidine, diphenyline and ortho-semidine),[122] confirming the duality of the acid dependence of the benzidine rearrangement.

The kinetic acidity dependence leads to the formulation of the possible constituent steps in eqns. 43-47 according to principles of acid-base catalysis:[138]

TABLE 6

Observed order in [H$^+$] in the rearrangement of hydrazoarenes

(Solvents: A = 60:40 dioxan:water; B = 95:5 ethanol:water; C = 75:25 acetone:water;
D = 25:75 methanol:water)

Compound R	(R'NHNHR') R'	Range in [H$^+$] M	Solvent	Order in [H$^+$]	Ref.
Ph	Ph	0.05 – 1	A	2	124,125
2-CH$_3$C$_6$H$_4$	2'-CH$_3$C$_6$H$_4$	0.01 – 0.5	A	1.3 – 2	126
2-CH$_3$OC$_6$H$_4$	Ph	0.002 – 0.3	A	1.1 – 2	127
2-FC$_6$H$_4$	2'-FC$_6$H$_4$	0.1 – 0.8	A	2	127
2-ClC$_6$H$_4$	2'-ClC$_6$H$_4$	0.8 – 2.8	A	2	127
2-BrC$_6$H$_4$	2'-BrC$_6$H$_4$	0.2 – 2.0	A	1.2 – 1.9	127
2-IC$_6$H$_4$	2'-IC$_6$H$_4$	0.7 – 1.6	A	1	127
2-PhC$_6$H$_4$	2'-PhC$_6$H$_4$	0.9 – 1.6	A	2	127
4-PhC$_6$H$_4$	Ph	0.004 – 0.6	A	2	128
4-PhC$_6$H$_4$	4'-PhC$_6$H$_4$	0.006 – 0.02	B	1.8	129
4-t-BuC$_6$H$_4$	4'-t-BuC$_6$H$_4$	0.01 – 0.05	B	1.9	130
α-Naphthyl	α-Naphthyl	5×10^{-6} – 0.03	A	1	131
β-Naphthyl	β-Naphthyl	0.01 – 0.15	A	1	132
α-Naphthyl	β-Naphthyl	0.01 – 0.2	A	1 – 1.2	133
β-Naphthyl	β-Naphthyl	0.001 – 0.02	C	1 – 2	134
α-Naphthyl	Ph	1×10^{-4} – 0.04	A	1 – 2	135
β-Naphthyl	Ph	8×10^{-4} – 0.6	A	1.1 – 2	136
Ph-NMe-NHPh		3×10^{-4} – 0.06	D	1.1 – 1.9	123
Ph-NAc-NHPh		2.6M HClO$_4$	H$_2$O	1	137

$$S + H^+ \xrightleftharpoons{\text{fast}} SH^+ \qquad\qquad (43)$$

$$SH^+ \xrightarrow{\text{slow}} \text{Products} \qquad\qquad (44)$$

$$SH^+ + H^+ \xrightleftharpoons{\text{fast}} SH_2^{2+} \qquad\qquad (45)$$

$$SH_2^{2+} \xrightarrow{\text{slow}} \text{Products} \qquad\qquad (46)$$

$$SH^+ + H^+ \xrightarrow{\text{slow}} \text{Products} \qquad\qquad (47)$$

The **first** protonation must certainly occur on nitrogen, as the most basic reaction site. Although the base strength of hydrazoarenes has not been measured experimentally, an estimate of the pK_a of hydrazobenzene has been given as ~ 0.[119,139] The rate of protonation of such a nitrogen base should be close to diffusion controlled;[140] hence the possibility that the first proton transfer could be slow is excluded for the one-proton process.

Proceeding to the second proton transfer step, the second nitrogen center is expected to be about 6 to 7 pK units less basic[139,141] (an estimate of $\Delta pK \sim 12$ has also been made[119,142]), so that there arises the possibility of a slow proton transfer to this nitrogen, in addition to another equilibrium protonation. The latter possibility has been looked upon as providing an attractive explanation for the reaction, the repulsion between the adjacent positive charges leading to rate-limiting scission of the N–N bond (eqn. 46). It is interesting to note for the situation that $[SH_2^{2+}]/[SH^+] \sim 10^{-7}$, the activation energy for decomposition of $[SH_2^{2+}]$ (eqn. 46) should be ~ 10 kcal/mole less than for decomposition of $[SH^+]$ (eqn. 44) in order that one- and two-proton mechanisms effectively compete with a given substrate.

Rate determining proton transfer (eqn. 47) should exhibit general acid catalysis, and equilibrium proton transfer specific acid catalysis. However, an early claim of buffer catalysis in the rearrangement of hydrazobenzene[143] was subsequently questioned and more stringent studies failed to confirm this type of behaviour.[144] Similarly, past usage of the Zucker-Hammett hypothesis[145,146] to distinguish between specific (log k $\propto$ -H_o) and general acid catalysis (log k $\propto$ log [H^+]) has been critically questioned in recent years.[147-149] The third approach, based on solvent isotope effects, forms part of the following section.

C. Hydrogen isotope effects

Whether or not proton transfer forms part of the rate-determining step can, in principle, be decided by means of hydrogen/deuterium kinetic isotope effects. In the problem at hand, we are concerned with the initial proton transfers between acid catalyst and substrate, as well as the eventual loss of the aromatic ring hydrogen atoms accompanying product formation. Information regarding these aspects is obtained from solvent and substrate kinetic isotope effects, respectively.

If the initial proton transfer were rate-determining, then replacing the catalyst H_3O^+ by D_3O^+ would cause a rate decrease[138,150]. If, on the other hand, a rapid reversible pre-equilibrium protonation were involved, then reaction would occur faster in the deuterated solvent. The latter result follows from the fact that D_3O^+/D_2O is a stronger acid than H_3O^+/H_2O, leading to a higher concentration of protonated substrate at a given stoichiometric acid concentration in the former case.

Studies with a number of substrates, characterized by reaction orders in [H^+] varying between 1 (e.g. 1,1'-hydra-zonaphthalene) and 2 (e.g. hydrazobenzene), yielded solvent isotope effect, k_{D_2O}/k_{H_2O}, values of between 2.1 and 4.8.[151] The average rate enhancement for the data in Table 7 is 2.0 ± 0.1 per added proton. While these results provided strong evidence that proton transfer occurred by fast pre-equilibrium processes in the above cases, studies with some other systems did not give concordant results. Thus in the following four cases the order in [H^+] was 2 but unexpectedly low solvent KIE values were found:[152] 2,2'-dibromo-

-hydrazobenzene, 2.0; 2,2'-dichlorohydrazobenzene, 2.3; 4-chlorohydrazobenzene, 3.0; N-acetylhydrazobenzene, 3.1. It was concluded by the authors that, for these compounds, the second protonation could become rate-determining, possibly as a result of the weak basicity caused by the electron with-drawing substituents. The solvent KIE for the second protonation was estimated as 0.7-1.0, indicating that this proton has been almost completely transferred in the transition state. It may be noted that the above results were obtained using 60% aqueous dioxan as solvent.

In another study,[139] but using purely aqueous solutions, it was found that hydrazo-o-toluene reacting by a one-proton mechanism yielded $k_{D_2O}/k_{H_2O} = 2.3$, but hydrazobenzene reacting by a two-proton mechanism yielded $k_{D_2O}/k_{H_2O} = 4.0$. Apparently the second protonation was associated with an isotope effect smaller than 2.3 (actually 1.7), which could once more be indicative of a slow second protonation and practically complete transfer of the proton in the transition state. This situation corresponds to the Bronsted coefficient α being close to unity, in which case general acid catalysis would not be observable.

The solvent isotope effect studies have thus shown that the first proton transfer occurs by a rapid pre-equilibrium, but the second proton transfer may occur either in a further pre-equilibrium or in a slow step depending on the substrate (and, perhaps, on a precise interpretation of the magnitude of the effect). However, the possibility then arises that the slow protonation may occur not on nitrogen but on aromatic carbon, and this will be discussed further subsequently.

Whether or not removal of a ring proton (vida supra) forms part of the rate determining step has been tested by means of a study of ring deuterated hydroazoarenes.[151] It has been found that deuterium substitution into hydrazobenzene at the 4,4'-positions, and at all positions but these, and into 1,1'-hydroazonaphthalene, at the 2,2'- and 4,4'-positions, did not affect the observed rate constants. Deprotonation must therefore be fast processes, occurring after the rate determining step.

TABLE 7

Kinetic Solvent Isotope Effects in Dioxan-H_2O and Dioxan-D_2O for Rearrangement of Hydrazoarenes.[151]

R	R'	$[H^+]$ or $[D^+]$	Order in H^+	k_{D_2O}/k_{H_2O} observed	k_{D_2O}/k_{H_2O} per added proton
1-Naphthyl	1-Naphthyl	0.010	1.0	2.3	2.3
2-Naphthyl	Phenyl	0.020	1.15	2.6	2.2
		0.31	1.75	3.8	2.1
ortho-Tolyl	ortho-Tolyl	0.010	1.25	2.1	1.8
		0.29	1.9	3.5	1.9
Phenyl	Phenyl	0.19	2.0	4.8	2.2

D. Nitrogen kinetic isotope effect

Although it had never been explicitly proven, the general assumption had been that scission of the N-N bond must be part of the rate-determining step. This assumption could no longer go unquestioned once evidence for rate-determining proton transfer was found in the two-proton process. If the second protonation were to occur on carbon, this would almost certainly require considerable activation energy which could then displace N-N bond scission as the rate limiting step. More-over, slow nitrogen protonation could allow the sequence shown in eqn. 48, which would be kinetically indistinguishable from eqns. 45 and 46 while entailing fast N-N bond breaking.

$$Ar\overset{+}{N}H_2NHAr \xrightarrow[slow]{HA} Ar\overset{+}{N}H_2\overset{+}{N}H_2Ar \xrightarrow{fast} Products \qquad (48)$$

The determination of nitrogen kinetic isotope effects, $k(^{14}N)/k(^{15}N)$, has often been used in the past to gain information on whether bond rupture to nitrogen is part of the rate determining step. For slow C-N or O-N bond rupture, nitrogen KIE values of 1.02 - 1.03 (i.e. 2-3%) have been observed and considered as diagnostic criteria of reaction mechanism.[153-158]

For rearrangement of hydrazobenzene in 75% aqueous
ethanol, k^{14}/k^{15} = 1.0203 ± 0.0007.[159] Since for a reaction
involving a rapid pre-equilibrium followed by the rate-
determining step (eqn. 49), the equilibrium isotope effect
would favour the ^{15}N species while the slow step favours the

$$A \overset{\text{fast}}{\rightleftharpoons} B \overset{\text{slow}}{\longrightarrow} C \qquad (49)$$

^{14}N species, the observed value of 1.023 actually represents
a lower limit of the KIE. Therefore there can be little doubt
that N-N bond rupture forms part of the rate-determining step.

E. Isotopic tracer studies

A number of isotopic studies have been made to probe the
question of inter-versus intramolecularity of the rearrangement
and all the results point to the latter situation.[160-163] In
one of the first studies of this type,[160] 2-[^{14}C]methylhydrazo-
benzene (103) was allowed to react in the presence of 2,2'-
dimethylhydrazobenzene (104) which in previous work has been
found to rearrange at comparable rate. Analysis of the
recovered 3,3'-dimethylbenzidine showed it to contain less than
0.03% of ^{14}C, thus eliminating the possibility of formation of
cross product (105) via an intermolecular pathway (eqn. 50):

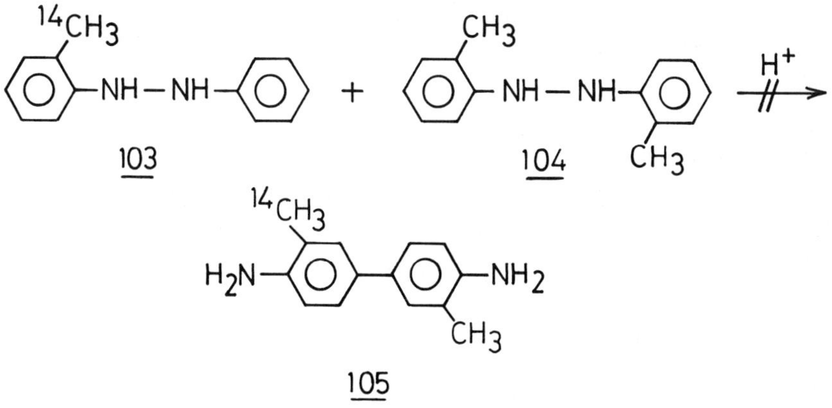

$$(50)$$

An interesting experiment of this type concerned the
rearrangement of N-acetylhydrazobenzene (106) in presence of
the decadeuterio analogue (107).[162] Mass spectral analysis of

the product showed parent peaks corresponding only to 108
and 109 with not a trace of the peak which would be derived
from the cross product.

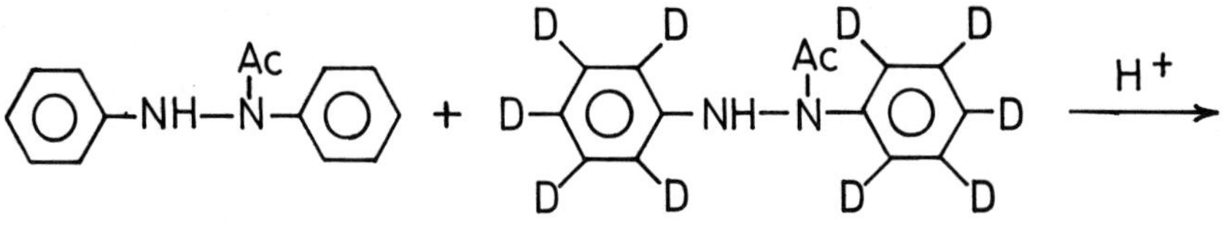

106 107

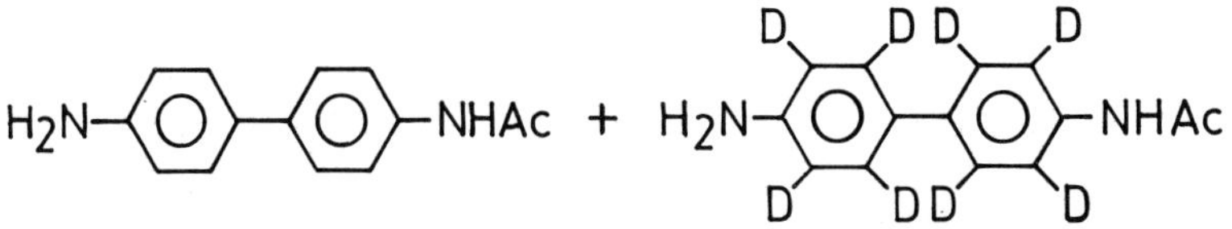

108 109 (51)

The proven intramolecularity of the benzidine rearrangement is in accord with some type of concerted mechanism in
which N-N bond rupture and C-C or C-N bond formation are
synchronous. Alternatively, it could be possible that the
substrate fragments but that the partners recombine very much
faster than they escape from the solvent cage into the bulk
solution (vide infra).

F. Theories of mechanism

As stated in Section XI. A, there are currently four
major theories concerning the mechanism of the benzidine
rearrangement. The salient features of these theories will now
be presented, emphasis being placed on the nature of transition states and intermediates. In a concluding section an
evaluation of the theories will be attempted, especially
regarding the quality of the agreement with results of kinetic
and isotopic criteria presented above.

1. The polar transition state theory

The central feature of this theory, advaced by Banthorpe,
Hughes and Ingold,[119,120,151] is the postulate that

both the one-proton and two-proton pathways entail the
formation of rate determining transition states of high
polarity due to partial heterolysis of the N-N bond in the
protonated substrate and the simultaneous development of the
incipient charges on the aromatic rings.   This is illustrated
in eqns. 52 and 53 which are applicable to the one-proton and
two-proton pathways, respectively:

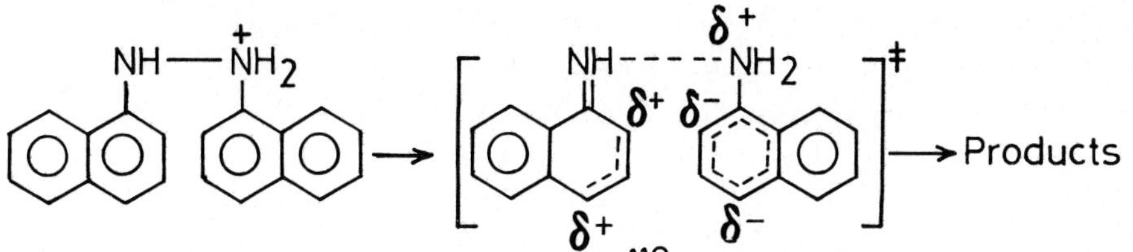

$$\text{(52)}$$

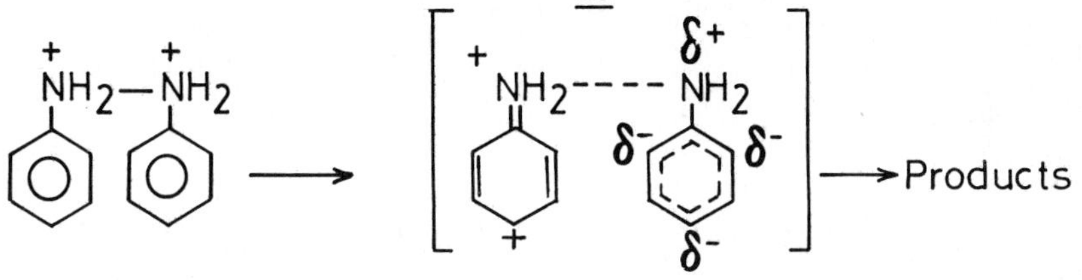

$$\text{(53)}$$

The charge distribution in the transition states entails
electrostatic interactions which lead to bonding modes
corresponding to product formation.   The different products
are then accounted for as follows:[119]   "When a family of
isomeric products is formed from a particular hydrazo-compound
the transition-states can be regarded as a family of cols,
all within a few kT of one another but on separated reaction
coordinates, the energy relationships of which can be
delineated by a study of the variation of isomer ratio with
temperature.   It is unlikely that the transition states
collapse directly to the finally isolated products.   The
differing sensitivities to medium of rate and products indicate
intermediates at a later stage than the transition state, and
the lack of a kinetic isotope-effect shows that the future
biaryl bond is established before either of the aromatic
hydrogens that have to be displaced are lost.   A family of σ-
bonded intermediates of quinonoid structure, analogous to

those occurring in aromatic S_E2 reactions, are likely; they lose protons to give benzenoid products in subsequent fast steps."

2. The π-complex theory

In this theory, which has been advocated by Dewar,[164] emphasis is placed on specific reaction intermediates, the π-complexes (112) and (113), which are postulated as being formed on rate-determining scission of the N-N bond in the one- and two-proton pathways (Scheme 5).

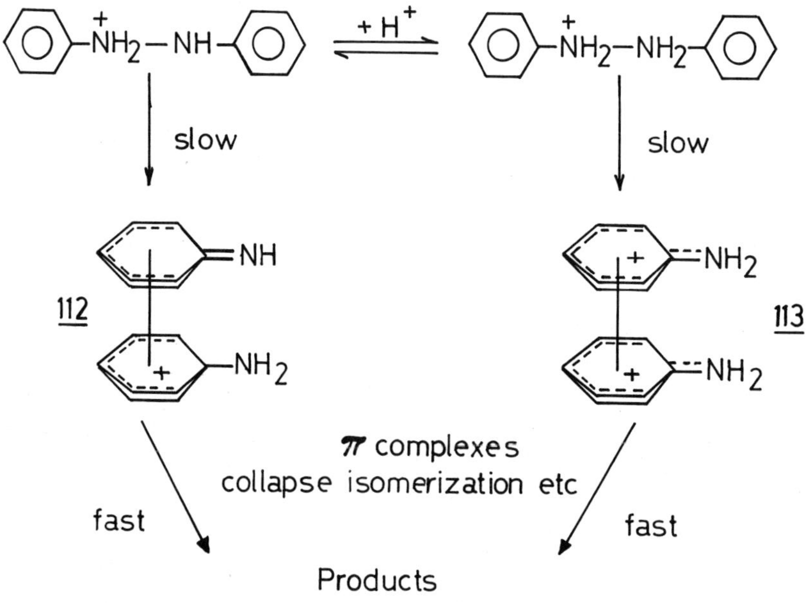

SCHEME 21

A consequence of the delocalized π-bonding which holds the rings in parallel planes, with the possibility of mutual rotation, is that the massive geometric changes required in product formation (especially of para-semidines) can readily be explained.

3. The caged-radical theory

The energetically favourable homolytic scission of an

N,N-diprotonated species into two cation radicals (eqn. 54) forms the basis of this proposal:

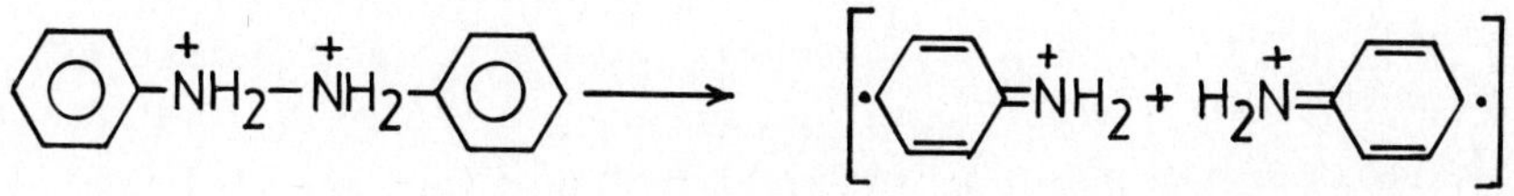

$$\left[\bigcirc\!\!-\!\!\overset{+}{N}H_2-\overset{+}{N}H_2-\!\!\bigcirc\right] \longrightarrow \left[\cdot\!\!\diamondsuit\!\!=\!\!\overset{+}{N}H_2 + H_2\overset{+}{N}\!\!=\!\!\diamondsuit\!\!\cdot\right] \qquad (54)$$

In order to account for intramolecularity of the overall process, the theory requires that the charged radicals be held within the solvent cage, without diffusing into the environment. The products are formed on recombination within the solvent cage by appropriate bonding modes.

Though numerous attempts to detect radicals in benzidine rearrangements by means of trapping experiments, as well as CIDNP, had been unsuccessful,[152,165] an important supporting study has been reported. In the course of the rearrangement of tetraphenylhydrazine in trifluoroacetic acid, the presence of the cation radical $Ph_2\overset{+\cdot}{N}H$ was revealed by its e.s.r. and u.v. spectra, which could be compared with the electrochemically generated authentic species.[166] Scheme 22 was proposed to account for these observations.

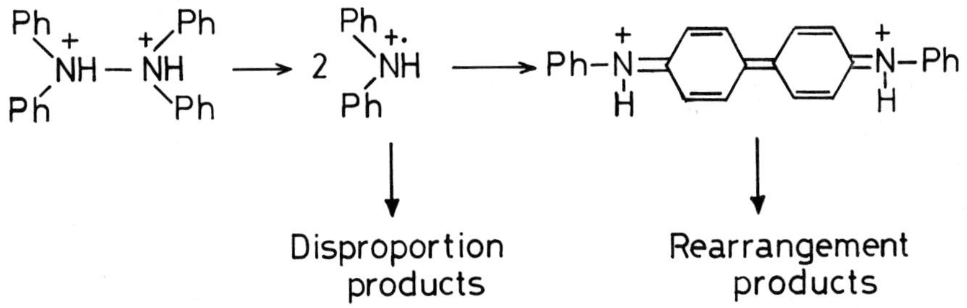

SCHEME 22

The formation of disproportionation products is readily accounted for by this scheme. The theory, however, cannot readily accommodate the one-proton mechanism since homolytic scission of the N-N bond would not be expected to be favourable in that case.

4. The C-protonated intermediate theory

First proposed in 1946[167] but receiving little support for the next 25 years, formation of reaction intermediates due to C-protonation has since been strongly advocated by a number of authors, including Olah, Lupes and others.[161,168-170] In this mechanism, the second protonation occurs not on nitrogen but on carbon to yield the key intermediate 114. Formation of the sp^3 center leads to a relatively flexible structure,

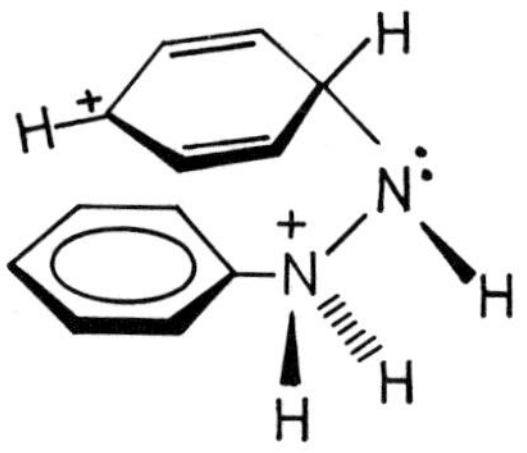

114

favourable for subsequent bonding modes. The products are derived from 114 by intramolecular alkylation pathways, as shown, for example, in Scheme 23 for benzidine formation. It is noteworthy that the species 115 could be observed by n.m.r. on treatment of hydrazobenzene with FSO_3H/SO_2ClF at $-78°$.

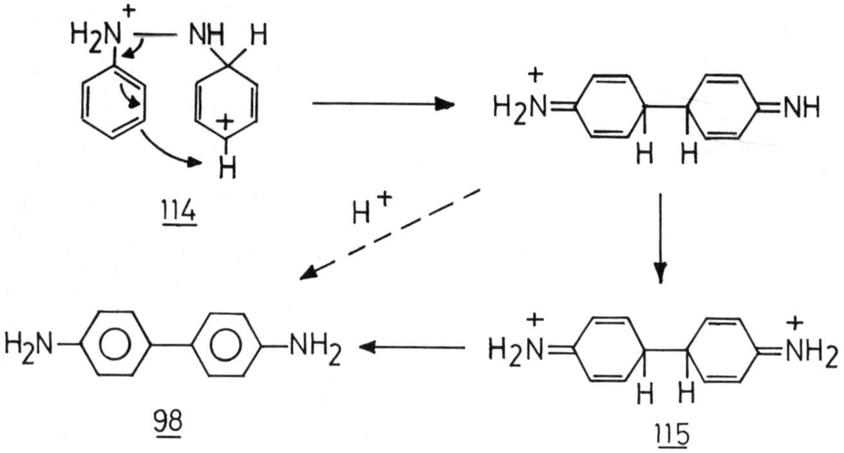

SCHEME 23

G. Summary and evaluation

Any acceptable mechanism of the benzidine rearrangement must in the first place be in agreement with the intra- molecularity requirement, as shown by isotopic and other studies related to the lack of formation of cross products. Of the four theories only the caged-radical proposal has any difficulty in meeting this condition. Though the complete lack, so far, of evidence for radicals which have escaped the solvent cage before recombination cannot be unheeded, yet there are other reactions involving cationic species where recombination of the reaction partners likewise is the dominant characteristic.[171-174] However, the caged radical theory appears to be limited conceptually to the two-proton mechanism, as has been noted above.

The polar transition state and π-complex theories appear to equally accommodate the principal kinetic findings. Although in fact both theories originally postulated two equilibrium protonations in accord with the earlier solvent KIE results, both can accommodate with but minor changes a second slow proton transfer as indicated by the later solvent KIE studies. Both theories comply with rupture of the N-N bond as part of the rate-determining step as shown by the recent $^{14}N/^{15}N$ KIE study, and the loss of aromatic hydrogens after the rate-determining step as shown by the lack of an effect of ring deuteration on the reaction rate. The apparent extreme emphasis of these two theories, one focussing on transition states and the other on intermediates, is note- worthy. Though the proponents of these theories consider them to be exclusive, perhaps this is not so after all. One can envisage the π-complexes to be formed from polar transition states, and while some structures of the π-complex type may be energy minima (intermediates) others may be energy maxima (transition states) along the reaction path.

The hypothesis of a C-protonated intermediate accommodated nicely the finding of a slow second proton transfer in some cases, since aromatic ring protonation is expected to be such. However, the dilemma arises that if ring protonation is the slow step, can N-N bond scission be part of the rate-

determining step as shown by the $^{14}N/^{15}N$ KIE? The schemes
invoking C-protonation have not, so far, incorporated this
possibility. If, on the other hand, the second proton is
transferred to nitrogen, one may still visualise this process
and N-N bond scission as occurring concertedly.

Given the complexity of the benzidine rearrangement
problem, it is probably not surprising that a complete under-
standing has not yet been reached. Though it is difficult at
this stage to predict which type of experiment may provide a
breakthrough, we suggest that further isotopic studies may in
fact well do so.

Considering the hypothesis of C-protonation in the first
instance, protonation at the 4- and 4'-positions of
hydrazobenzene may well be more probable than at the 1- or 1'-
positions. However, there has been no attempt to detect
deuterium (tritium) exchange into hydrazobenzene recovered
from partial reaction in deuterated (tritiated) medium.
(Olah and co-workers[168] examined the benzidine product
obtained from reaction in D_2SO_4, but this of course cannot
yield information on exchange at the 4,4'-positions.)

More important, perhaps, would be to evaluate the $^{14}N/$
^{15}N isotope effect for a reaction proceeding via a one-proton
mechanism, and to differentiate between processes in which the
second proton transfer occurs in a pre-equilibrium as
opposed to a slow step, via the nitrogen isotope effect.

Finally, one can point to the absence, so far, of any
reported study of $^{12}C/^{14}C$ (or $^{12}C/^{13}C$) kinetic isotope effects.
Such results would be informative concerning the question
whether C-C bond making is part of the rate determining step,
especially in the formation of diphenylines and the ortho-
and para-semidines.

XII   REARRANGEMENT OF DIAZONIUM SALTS

The possibility of rearrangement of nitrogen atoms in
the diazonium cation (eqn. 55) is of considerable intrinsic
interest. The process also has bearing on mechanistic studies

$$[Ar-^{15}N{\equiv}N]^+ \rightleftharpoons [Ar-N{\equiv}^{15}N]^+ \qquad (55)$$

with the azoxy, azo and hydrazo series since such studies usually implicitly assume that the above rearrangement does not in fact occur. For example, in the preparation of ^{15}N labelled compounds via diazonium salts, or in their degradation, it is normally assumed that the integrity of the isotopic label remains undisturbed.[175]

If the rearrangement in eqn. 55 could be demonstrated to occur, one would wish to elucidate its mechanism. For example, is it a unitary process occurring via a single transition state, or does it involve the formation of some discrete intermediate? If the former, what is the nature of the transition state? If the latter, could the intermediate be detectable by kinetic or perhaps even by spectral techniques?

Another aspect of the problem relates to the reaction conditions under which the isomerisation may occur. For example, if it occurs in aqueous solution, what relation does it bear to the competing hydrolysis in which phenol is produced? Is there a common intermediate to these competing processes? The result of these considerations is that the mechanism of solvolysis, or of nucleophilic substitution in general, of diazonium salts becomes intimately involved with the N_α-N_β rearrangement.

In the discussion which follows, the subject matter is restricted to heterolytic processes, i.e. homolytic or photolytic reactions of diazonium salts are excluded from consideration (see refs. 176,177).

It is interesting that the idea of isotopic scrambling in the diazonium ion originated from a kinetic study. An increase in the rate of hydrolysis of PhN_2^+ by added thiocyanate ion suggested that an intermediate was formed which could revert to the diazonium ion.[178] Possible structures for the intermediate which were proposed included the spiro diazirine cation 116 with equivalent nitrogens and the caged pair 117 in which the nitrogens are non-equivalent. In either case isotopic rearrangement should accompany solvolysis. It took a number of years, however, and work in several laboratories before the feasibility of this reaction (eqn. 55) become fully established, probably because the magnitude of the phenomenon

116   117

under commonly used conditions is quite small.

It was initially reported[179] that the hydrolysis of $[Ph-^{15}N\equiv N]^+ BF_4^-$ in water at 35°C is accompanied by a small degree of rearrangement, i.e. $k_{rear}/k_{solv} = 0.014$. Although a subsequent study[180] failed to confirm this result, more extensive work from the original laboratory showed that several substituted benzenediazonium salts follow this type of process.[181] The rearrangement : solvolysis rate ratio for the compounds examined was found to be as follows: p-CH$_3$, 0.030; m-CH$_3$, 0.018; p-OCH$_3$, 0.038; p-Cl, 0.023. It is interesting that not only was there no particular trend observed with change in substituent (i.e. no apparent correlation with Hammett σ or with migratory aptitudes), but the rate ratios were practically independent of temperature (36° to 62°C). The rearrangement was not suppressed by the addition of nucleophilic species in high concentration (2.9 $\underline{M}$ NaBr). Also, extensive studies of salt effects on the hydrolysis brought the earlier result on the effect of SCN$^-$ into the range of normal salt effects in these systems. The spiro structure (116) was abandoned at this point as an intermediate in the isomerisation of $[Ph-^{15}N\equiv N]^+$.[182]

The lack of a variable which would influence the rearrange ment hampered further advances for several years, until it was discovered that solvent has a remarkable effect on these processes. The data in Table 8[183] show that, by varying the solvent, k_r/k_s values as large as 0.09 can be obtained. It is interesting that, comparing the data for 2,2,2-tri-fluoroethanol (TFE) and water, the overall increase in the k_r/k_s value derives from an 8-fold increase in k_r and only 1.5-fold increase in k_s. The manner in which these results were obtained can be given briefly. $[Ph-\overset{+}{N}\equiv^{15}N]BF_4^-$ obtained by diazotization of aniline with Na$^{15}NO_2$ (99.2% ^{15}N) is allowed

TABLE 8

Isotopic Rearrangement of Substituted Benzenediazonium-β-^{15}N Tetrafluoroborate[183]

Conditions	Substituent	P (%)[a]	k_r/k_s	$10^4 \, k_s$ (s^{-1})	$10^6 \, k_r$ (s^{-1})
H_2O, 30°C	H	1.89	0.016	0.927	1.48
70 wt % TFE, 30°C	H	7.14	0.064	1.581	10.1
85 wt % TFE, 30°C	H	7.65	0.069	1.642	11.3
97 wt % TFE, 30°C	H	8.16	0.074	1.607	11.9
TFE, 30°C	H	7.96	0.072	1.590	11.4
TFE, 5°C	H	8.66	0.079	0.023	0.18
TFE, 30°C	2,4,6-D	6.41	0.057	1.078	6.14
TFE, 30°C	H	9.05	0.083	0.778	6.46
1 M H_2SO_4, 35°C	H	1.89	0.016	1.67	2.67
85% H_3PO_4, 30°C	H	9.45	0.087	0.524	4.56
TFE, 64°C	4-OCH$_3$	8.28	0.075	0.026	0.19
TFE, 40°C	4-CH$_3$	8.94	0.082	0.647	5.31

[a]Percent isotopic rearrangement adjusted to exactly 70% dediazoniation.

to react to ~70% completion, at which point the reaction is quenched by addition of 2-naphthol. The azo compound formed from any unreacted diazonium salt is reduced with dithionite to aniline, which is directly analysed mass spectrometrically.

An even greater degree of rearrangement was observed when hexafluoro-2-propanol (HFIP) was used as solvent in place of TFE, and when 2,4,6-trimethylbenzenediazonium ion was used in place of PhN$_2^+$ (Table 9).[184] These findings appear to reflect the influence of steric crowding at the reaction site, noting the HFIP has nucleophilicity slightly lower, and ionizing power slightly higher than TFE, and that use of the much better ionizing FSO$_3$H solvent leads to less isotopic scrambling compared to HFIP.

Parallel to the studies on nitrogen rearrangement,

TABLE 9

Extent of N_α-N_β Rearrangement of Diazonium Salts Accompanying
70% Dediazoniation at 25°C, 1 atm N_2.[184]

Diazonium Salt[a]	Solvent	P,%[b]	k_r/k_s[b]	$10^5 k_r$, s^{-1}
1	TFE	7.96	0.072	0.581
1	HFIP	10.47	0.098	0.776
1	FSO$_3$H	3.66	0.032	0.149
2	TFE	20.89	0.225	3.72
2	HFIP	36.97	0.559	10.8
2	FSO$_3$H	15.62	0.157	3.72

[a] 1 = benzenediazonium fluoroborate; 2 = 2,4,6-trimethyl-
benzene diazonium fluoroborate

[b] Percent isotopic rearrangement adjusted to exactly 70%
dediazoniation

Zollinger and co-workers have also investigated isotopic
exchange between diazonium salts and external nitrogen.
Results obtained for TFE as solvent are given in Table 10,[183]
corresponding to ^{15}N labelled diazonium fluoroborate reacting
to 70% completion under various pressures of isotopically
normal N_2 (i.e.~0.4% ^{15}N). It is seen that as the pressure
of N_2 is increased, the ^{15}N content of the unreacted
diazonium salt steadily decreases, indicative of exchange with
external N_2 (eqn. 56):

$$Ar-\overset{*}{N}_2^+ + N_2 \rightleftharpoons Ar-N_2^+ + \overset{*}{N}_2 \qquad (56)$$

Changing the solvent to HFIP, or use of 2,4,6-trimethyl-
benzenediazonium ion, led to even larger degrees of exchange
(Table 11).[184] This was further corroborated by determining
the effect of varying the partial pressure of N_2 (in mixtures
with argon) on the dediazoniation rate, N_α-N_β rearrangement,
and exchange with external N_2.[185]

TABLE 10

Exchange of <u>Para</u>-Substituted β-^{15}N-Diazonium Ions with
External N_2 in trifluoroethanol.[183]

Substituent	% Dediazoniation	Conditions	$^{15}N\equiv^{14}N$ Content of azo dye (%)
H	0	1 atm,air,25°C	99.20[a]
H	70.4	1 atm,air,25°C	98.60 ± 0.44[b]
H	73.2	20 atm,$^{14}N_2$,25°C	98.23 ± 0.48[b]
H	69.9	300 atm, $^{14}N_2$,25°C	96.23 ± 0.47[b]
H	70.1	300 atm,$^{14}N_2$, 25°C	96.99 ± 0.26[b]
H	72.5	300 atm,$^{14}N_2$, 25°C	96.89 ± 0.27[b]
H	62.5	1000 atm,$^{14}N_2$, 25°C	94.71 ± 0.43[b]
NO_2	0	1 atm,air, RT	99.28[a]
NO_2	65.4	315 atm,$^{14}N_2$,64.0°C	98.33 ± 0.23[b]
OCH_3	0	1 atm,air, RT	98.87[a]
OCH_3	67.3	310 atm,$^{14}N_2$,64.0°C	97.55 ± 0.44[b]

[a]Reference standard

[b]95% confidence limits

These studies have provided the first evidence for the reaction between N_2 and a purely organic reagent in solution. Moreover, a parallel experiment involved the incorporation of CO; reaction of $PhN_2^+BF_4^-$ in TFE under 320 atm of carbon monoxide gave 5.2% of 2,2,2-trifluoroethyl benzoate.[183]

What mechanistic conclusions can be drawn from these studies? First, it is extremely unlikely that solvolysis, N_α-N_β rearrangement, nitrogen exchange and carbonylation could occur via a single transition state. Moreover, one must critically examine whether a mechanism involving a single

TABLE 11

Solvent Effects on the Rearrangement and the Exchange Reaction
of Diazonium Salts.[184]

Solvent	Reaction[a]	Salt 2	Salt 1	2/1[b]
HFIP	Re	36.97	10.47	3.53
	Ex	16.50	6.26	2.64
	Re/Ex	2.24	1.67	
TFE	Re	20.89	7.96	2.62
	Ex	6.33	2.46	2.57
	Re/Ex	3.30	3.24	
FSO_3H	Re	15.62	3.33	4.69
HFIP[c]	Re	1.77	1.32	
TFE	Ex	1.61	1.57	

[a] Re = N_α - N_β rearrangement. Ex = percent exchange with
external nitrogen accompanying 70% dediazoniation.

[b] Ratio of corresponding values for diazonium salts 2 and 1
respectively (see Table 9).

[c] Ratio of the corresponding values in HFIP and TFE,
respectively. The ratio for the exchange reaction was
obtained by multiplying the observed ratio by 0.62 so that
the difference in N_2 concentration in TFE and HFIP is
compensated for.

intermediate suffices, i.e. rate-limiting formation of a phenyl
cation followed by its competitive reaction with solvent, the
β-atom of the N_2 molecule that is split off, external N_2, or
CO molecules. If rearrangement and exchange were actually
occurring through the same intermediate, then the two processes
should occur to an equal extent, as it can reasonably be
inferred that in the TFE as well as HFIP systems the first

solvation shell would contain at least one external N_2 molecule. However, the data summarised in Table 11[184] show that the amount of rearrangement is 1.67-3.30 greater than exchange.

The results thus imply that not one but two intermediates are reversibly formed in the dediazoniation of ArN_2^+ in TFE and HFIP, and that the intermediate for rearrangement must occur earlier on the reaction coordinate than that for exchange. It is proposed that the first intermediate is a tight nitrogen-aryl cation molecule-ion pair, while the second intermediate is the solvent separated molecule-ion pair or the free aryl cation. Dediazoniation products can be formed by nucleophilic attack on both intermediates as shown in Scheme 24, which is analogous to that first postulated by Winstein for solvolytic processes.[186]

$$Ar-N_2^+ \rightleftharpoons [Ar^+N_2] \rightleftharpoons Ar^+ + N_2$$

$$\downarrow Nu \qquad \downarrow Nu$$

Products   Products

SCHEME 24

The above studies have shown elegantly how, through the influence of solvent and steric effects, the barely detectable rearrangement of benzenediazonium cation in water could become an important pathway in TFE or HFIP, being even more accentuated in the case of substitution at the aryl 2- and 6-positions. A statistical evaluation of the kinetic data demonstrated recently[186a] that Scheme 24 is the only mechanism which fits the data on a 99% confidence level and which is chemically reasonable.

The next stage of such studies could be to stabilize reaction intermediates sufficiently so as to be detectable spectroscopically. This has not yet been found possible in solvolytic type processes of diazonium salts. However the UV irradiation (in polymer films or 9M LiCl aqueous glass at $77^{\circ}K$) of arenediazonium salts substituted at the 4-position by dialkylamino groups gives rise to e.s.r. signals which have been assigned to the aryl cation in the triplet state.[187] This study provides conclusive evidence for the existence of these Ar^+ cations.

Parallel to the experimental studies, highly informative theoretical studies have come forth related to the benzene-diazonium ion and its rearrangement.[188-190] The latest of these studies[190] suggests that the dissociation-recombination mechanism for rearrangement could be energetically competitive with a concerted, intramolecular, process. The calculated energy profile (Figure 2) shows a meta-stable bridged inter-mediate in a high-energy shallow potential well. The trans-ision states for formation and decomposition of the intermediate are asymmetric, one of the nitrogens being weakly bound to the C-1 ring carbon and the other not at all.

Figure 2. Schematic energy profile for N_α-N_β rearrangement in the benzenediazonium ion.[190]

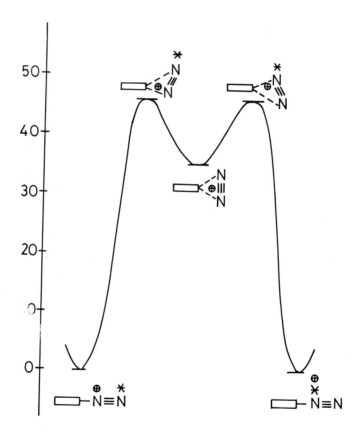

Reactivity studies of systems involving aryl cations, e.g. as formed by dediazoniation of ArN_2^+ in nucleophilic systems, will no doubt continue in the future. The high reactivity and low selectivity of Ar^+ towards nucleophiles can lead to apparently conflicting results, indicative of both unimolecular and bimolecular mechanisms.[191,192] Kinetic isotope effect results ($^{14}N/^{15}N$ and secondary aromatic hydrogen KIE's) for the hydrolysis of benzenediazonium ion have been interpreted on the basis of a unimolecular mechanism.[193,194] Thus the overall goal which we set at the beginning of this section has yet to be attained, though occurrence of the rearrangement has been fully established.[195]

REFERENCES

1   E. Bamberger and K. Landsteiner, Ber., 26 (1893) 482;
     E. Bamberger, ibid, 30 (1897) 1248.

2   A.F. Holleman, J.C. Hartogs and T. van der Linden, Ber., 44 (1911) 704.

3   E.D. Hughes and G.T. Jones, J. Chem. Soc., (1950) 2678.

4   S. Brownstein, C.A. Bunton and E.D. Hughes, J. Chem. Soc., (1958) 4354.

5   D.V. Banthorpe, E.D. Hughes and D.L.H. Williams, J. Chem. Soc.,
         (1964) 5349.

6   D.V. Banthorpe, J.A. Thomas and D.L.H. Williams, J. Chem. Soc.,
         (1965) 6135.

7   B.A. Geller and L.N. Dubrova, J. Gen. Chem. USSR, 30 (1960) 2627.

8   W.N. White and J.T. Golden, Chem. and Ind., (1962) 138.

9   W.N. White et al., J. Org. Chem., 35 (1970) 737, 965, 1803, 2048;
         J. Am. Chem. Soc., 86 (1964) 1517.

10   W.N. White in Mechanisms of Molecular Migration, Ed. B.S. Thyagarajan,
         Vol. 3, Wiley Interscience, New York, 1971, p.109.

11   B. Capon, M.J. Perkins and C.W. Rees, Organic Reaction Mechanisms,
         (1965), p.176, Interscience.

12   D.V. Banthorpe and J.A. Thomas, J. Chem. Soc., (1965) 7149.

13   W.N. White, J.R. Klink, D. Lazdins, C. Hathaway, J.T. Golden and
         H.S. White, J. Am. Chem. Soc., 83 (1961) 2024.

14   D.V. Banthorpe and J.G. Winter, J.C.S. Perkin II, (1972) 1259.

15   Z.J. Allan, Tetrahedron Lett., (1971) 4225.

16   B.A. Geller and L.S. Samsvat, Zh. Obshch. Khim., 34 (1964) 613.

17   E. Bamberger, Ber., 27 (1894) 584; 28 (1895) 399.

18   S.R. Hartshorn and J.H. Ridd, J. Chem. Soc. B., (1968) 1063.

19   J.H. Ridd and E.F.V. Scriven, J.C.S. Chem. Comm., (1972) 641.

20   O. Fischer and E. Hepp, Ber., 19 (1886) 2991; 20 (1887) 1247, 2471,
         2479.

21   P.W. Neber and H. Raucher, Ann., 550 (1942) 182.

22   J. Houben, Ber., 46 (1913) 3984.

23   C.K. Ingold, Structure and Mechanism in Organic Chemistry (2nd
         edition) p.901, Bell, London (1969); H.J. Shine, Aromatic
         Rearrangements pp. 231-5, Elsevier, Amsterdam (1967).

24   M.J.S. Dewar, Molecular Rearrangements (Ed. P. de Mayo), p.310,
         Interscience, New York (1963).

25   W. Macmillen and T.H. Reade, J. Chem. Soc., (1929) 585.

26   T.D.B. Morgan, D.L.H. Williams and J.A. Wilson, J.C.S. Perkin II,
         (1973) 473.

27   D.L.H. Williams, Tetrahedron, 31 (1975) 1343; Int. J. Chem. Kin., 7
          (1975) 215.

28   D.L.H. Williams, J.C.S. Perkin II, (1975) 655.

29   T.D.B. Morgan and D.L.H. Williams, J.C.S. Perkin II, (1972) 74.

30   I.D. Biggs and D.L.H. Williams, J.C.S. Perkin II, (1975) 107.

31   K.M. Ibne-Rasa, J. Am. Chem. Soc., 84 (1962) 4962.

32   B.C. Challis and R.J. Higgins, J.C.S. Perkin II, (1972) 2365;
          B.C. Challis, R.J. Higgins and A.J. Lawson, ibid., p.1831.

33   Z. Vrba and Z.J. Allan, Tetrahedron Lett., (1968) 4507.

34   Z.J. Allan, Tetrahedron Lett., (1971) 4225.

35   E. Bamberger, Ann. Chem., 424 (1921) 233, 297; 441 (1925) 207.

36   S. Okazaki and M. Okumura, unpublished work, quoted by S. Oae,
          T. Fukomoto and M. Yamagami, Bull. Chem. Soc. Japan, 34 (1961)
          1873; 36 (1963) 601.

37   I.I. Kukhtenko, Zhur. org. Khim., 7 (1971) 324 (English translation
          edition).

38   H.E. Heller, E.D. Hughes and C.K. Ingold, Nature, 168 (1951) 909.

39   G. Kohnstam, W.A. Petch and D.L.H. Williams, to be published.

40   P.G. Gassman, Acc. Chem. Res., 3 (1970) 26.

41   G.T. Tissue, M. Grassmann and L.W. Lwowski, Tetrahedron, 24 (1968)
          999; W.E. Truce, J.W. Fieldhouse, D.J. Vrenair, J.R. Norell,
          R.W. Campbell and D.G. Brady, J. Org. Chem., 34 (1969) 3097.

42   D. Gutshke and A. Heesing, Ber., 106 (1973) 2379.

43   G. Bender, Ber., 19 (1886) 2272.

44   E.D. Hughes and C.K. Ingold, Quart. Rev., 6 (1952) 34.

45   A.E. Bradfield, K.J.P. Orton and I.C. Roberts, J. Chem. Soc.,
          (1928) 782. M. Richardson and F.G. Soper, J. Chem. Soc.,
          (1929) 1873.

46   K.J.P. Orton and A.E. Bradfield, J. Chem. Soc., (1927) 986; C. Beard
          and W.J. Hickinbottom, ibid., (1958) 2982.

47   E. Slosson, Ber., 28 (1895) 3265.

48   A.R. Olson, G.W. Porter, F.A. Long and R.S. Halford, J. Am. Chem.
          Soc., 58 (1936) 2467.

49   R.P. Bell, Proc. Roy. Soc. (London), Ser. A., 143 (1934) 377;
          R.P. Bell and R.V.H. Levinge, ibid., 151 (1935) 211; R.P. Bell,
          J. Chem. Soc., (1936) 1154.

50   J.M.W. Scott and J.C. Martin, Can. J. Chem., 43 (1965) 732.

51   M.J.S. Dewar in Theoretical Organic Chemistry, Butterworths, London,
          1959, pp.185-7.

52   E. Bamberger and E. Hindermann, Ber., 30 (1897) 654.

53   G. Illuminati, J. Am. Chem. Soc., 78 (1956) 2603.

54   W.J. Spillane and F.L. Scott, J. Chem. Soc. (B), (1968) 779.

55   Z. Vrba and Z.J. Allan, Tetrahedron Letters, (1968) 4507;
          Coll. Czech, Chem. Comm., 33 (1968) 2502.

56   W.J. Spillane, F.L. Scott and C.B. Goggin, J. Chem. Soc. (B), (1971),
          2409.

57   E.A. Shilov, M.N. Bogdanov and A.E. Shilov, Doklady Akad. Nauk
          S.S.S.R., 92 (1953) 93.

58   P.B. Fischer and H. Zollinger, Helv. Chim. Acta., 53 (1970), 1306.

59   A.H. Blatt, Organic Reactions, Vol. 1, Wiley, New York, 1942, p.342.

60   K. Rosenmund and W. Schnurr, Ann. Chem., 460 (1928) 56.

61   R. Martin and J.M. Betoux, Bull. Soc. Chim. France, (1969) 2079.

62   Y. Ogata and H. Tabuchi, Tetrahedron, 20 (1964) 1661.

63   F. Krausz and R. Martin, Bull. Soc. Chim. France, (1965) 2192;
          C.R. Hauser and E.H. Man, J. Org. Chem., 17 (1952) 390.

64   N.M. Cullinane, R.A. Woolhouse and B.F.R. Edwards, J. Chem. Soc.,
          (1961) 3842, and earlier papers.

65   H.J. Shine, Aromatic Rearrangements, p.81, Elsevier, Amsterdam (1967).

66   H. Hart and R.J. Elia, J. Am. Chem. Soc., 76 (1954) 3031.

67   M.M. Sprung and E.S. Wallis, J. Am. Chem. Soc., 56 (1934) 1715;
          W.I. Gilbert and E.S. Wallis, J. Org. Chem., 5 (1940) 184.

68   M.J.S. Dewar and N.A. Puttnam, J. Chem. Soc., (1960) 959 and earlier
          papers.

69   N.M. Cullinane, R.A. Woolhouse and G.B. Carter, J. Chem. Soc.,
          (1962) 2995.

70   M.J.S. Dewar and P.A. Spanninger, J.C.S. Perkin II, (1972) 1204.

71   P.A. Spanninger and J.L. Von Rosenberg, J. Am. Chem. Soc., 94 (1972)
          1973.

72   W. Haegele and H. Schmid, Helv. Chim. Acta, 41 (1958) 657 and
          earlier papers.

73   J. Borgulya, R. Madeja, P. Fahrini, H.J. Hansen, H. Schmid and
          R. Barner, Helv. Chim. Acta, 56 (1973) 14.

74   D.A. McCauley and A.P. Lien, J. Am. Chem. Soc., 74 (1952) 6246;
          M. Kilpatrick, J.A.S. Bett and M.L. Kilpatrick, J. Am. Chem.
          Soc., 85 (1963) 1038.

75   See L.I. Smith, Org. Reactions, 1 (1942) 370.

76   H. Steinberg and F.L.J. Sixma, Rec. Trav. Chim., 81 (1962) 185;
        V.A. Koptyug, I.S. Isaev and N.N. Vorozhtsov, Proc. Akad. Sci.
        U.S.S.R., 149 (1963) 191.

77   N.N. Vorozhtsov and V.A. Koptyug, J. Gen. Chem. U.S.S.R., 30
        (1960) 1014.

78   G.A. Olah, J. Am. Chem. Soc., 87 (1965) 1103.

79   R.H. Allen, J. Am. Chem. Soc., 82 (1960) 4856.

80   E. Unseren and A.P. Wolf, J. Org. Chem., 27 (1962) 1509.

81   G.M. Moore and A.P. Wolf, J. Org. Chem., 31 (1966) 1106.

82   D.A. McCauley and A.P. Lien, J. Am. Chem Soc., 74 (1952) 6246.

83   A. Streitwieser and L. Reif, J. Am. Chem. Soc., 86 (1964) 1988.

84   O. Wallach and L. Belli, Chem. Ber., 13 (1880) 525.

85   G.G. Spence, E.C. Taylor and O. Buchardt, Chem. Rev., 70 (1970) 231.

86   D.J.W. Goon, N.G. Murray, J.P. Schoch, and N.J. Bunce, Can. J. Chem.,
        51 (1973) 3827.

87   A. Angeli, Gazz. Chim. Ital., 46, II (1916) 67.

88   C.S. Hahn and H.H. Jaffé, J. Am. Chem. Soc., 84 (1962) 946.

89   P.H. Gore and G.K. Hughes, Australian J. Sci. Res., 3A (1950) 136
        ibid; 4A (1951) 185.

90   B.T. Newbold, J. Chem. Soc., 1965, 6972; B.T. Newbold, private
        communication; M.H. Akhtar, M.Sc. Thesis, University of Moncton
        (1967).

91   M.M. Shemyakin, V.I. Maimind, and B.K. Vaichunaite, Zh. Obshch. Khim.,
        28 (1958) 1708; Chem. Abstr., 53 (1959) 1201; Chem. Ind.
        (London), (1958) 755; Izv. Akad. Nauk. SSSR, Otdel Khim. Nauk,
        (1960) 866; Chem. Abstr., 54 (1960) 24474.

92   L.C. Behr and E.C. Hendley, J. Org. Chem., 31 (1966) 2715.

93   S. Oae, T. Fukumoto, and M. Yamagami, Bull. Chem. Soc. Japan, 36
        (1963) 601; ibid., 34 (1961) 1873.

94   M.M. Shemyakin, T.E. Agadzhanyan, V.I. Maimind, and R.V. Kudryavtsev,
        Izv. Akad. Nauk SSSR, Ser. Khim., (1963) 1339; Chem. Abstr., 59
        (1963) 12619; M.M. Shemyakin, T.E. Agadzhanyan, V.I. Maimind,
        R.V. Kudryavtsev, and D.N. Kursanov, Dokl. Akad. Nauk SSSR, 135
        (1960) 346; Chem. Abstr., 55 (1961) 11337.

95   M.M. Shemyakin and V.I. Maimind, in "Recent Progress in the Chemistry
        of Natural and Synthetic Coloring Matters and Related Fields",
        T.S. Gore, Ed., Academic Press, New York, 1962.

226

96   J.F. Bunnett, E. Buncel, and K.V. Nahabedian, J. Am. Chem. Soc., 84
(1962) 4136.

97   P.H. Gore, Chem. Ind. (London), (1959) 191.

98   E. Buncel, in "Mechanisms of Molecular Migrations", Vol. 1, B.S.
Thyagarajan Ed., Wiley, New York, 1968.

99   E. Buncel and B.T. Lawton, Chem. Ind. (London), (1963) 1835.

100   E. Buncel and B.T. Lawton, Can. J. Chem., 43 (1965) 862;
E. Buncel and W.M.J. Strachan, Can. J. Chem., 48 (1970) 377.

101   E. Buncel and W.M.J. Strachan, Can. J. Chem., 47 (1969) 911.

102   H.H. Jaffé and R.W. Gardner, J. Am. Chem. Soc., 80 (1958) 319;
S.J. Yeh and H.H. Jaffé, J. Am. Chem. Soc., 81 (1959) 3274.

103   E. Buncel, Acc. Chem. Res., 8 (1975) 132.

104   R.A. Cox, J. Am. Chem. Soc., 96 (1975) 1059.

105   E. Buncel, A. Dolenko, I.G. Csizmadia, J. Pincock, and K. Yates
Tetrahadron, 24 (1968) 6671.

106   G.A. Olah, K. Dunne, D.P. Kelly and K.Y. Mo, J. Am. Chem. Soc.,
94 (1972) 7438.

107   C.S. Hahn, K.W. Lee, and H.H. Jaffé, J. Am. Chem. Soc., 89 (1967)
4975.

108   D. Duffey and E.C. Hendley, J. Org. Chem. 33 (1968) 1918.

109   D. Duffey and E.C. Hendley, J. Org. Chem. 35 (1970) 3579.

110   A. Dolenko and E. Buncel, Can. J. Chem. 52 (1974) 623; R.A. Cox,
A. Dolenko and E. Buncel, J. Chem. Soc., Perkin Trans II,
(1975) 471.

111   R.A. Cox and E. Buncel, Can. J. Chem. 51 (1973) 3143; J. Am. Chem.
Soc., 97 (1975) 1871.

112   S. Oae and J. Maeda, Tetrahedron, 28 (1972) 2127.

113   S. Oae, T. Maeda, S. Kozuka, and M. Nakai, Bull. Chem. Soc. Japan,
44 (1971) 2495.

114   A.W. Hoffmann, Proc. Roy. Soc., 12 (1863) 576.

115   H. Schmidt and G. Schultz, Chem. Ber., 11 (1878) 1754.

116   M. Vecera, J. Gasparic and J. Petranek, Coll. Czech. Chem. Commun.,
23 (1958) 249.

117   Z.J. Allan, Liebigs Ann. Chem., (1978) 705; Monatsch. Chem. 106
(1975) 429.

118   H.J. Shine, in "Mechanisms of Molecular Migrations", (Ed. B.S.
Thyagarajan), Vol. 2, Interscience, New York, 1969, p.191;
"Aromatic Rearrangements", Elsevier, New York, 1967, p.126.

119   D.V. Banthorpe, Topics Carbocyclic Chem., 1 (1969) 1.

120   C.K. Ingold, Chem. Soc. Spec. Publ., No. 16 (1962) 118.

121   G.S. Hammond and H.J. Shine, J. Am. Chem. Soc., 72 (1950) 220.

122   R.B. Carlin and R.C. Odioso, J. Am. Chem. Soc., 76 (1954) 2345.

123   W.N. White and R. Preisman, Chem. Ind. (London), (1961) 1752.

124   C.A. Bunton, C.K. Ingold and M. Mhala, J. Chem. Soc., (1957) 1906.

125   D.V. Banthorpe, E.D. Hughes, C.K. Ingold and J. Roy, J. Chem. Soc.,
          (1962) 3294.

126   D.V. Banthorpe and E.D. Hughes, J. Chem. Soc., (1962) 3314.

127   D.V. Banthorpe, A. Cooper and C.K. Ingold, Nature (London), 216
          (1967) 232.

128   D.V. Banthorpe, A. Cooper and M. O'Sullivan, J. Chem. Soc., B, (1971)
          2054.

129   H.J. Shine and J.P. Stanley, Chem. Commun., (1965) 294; J. Org.
          Chem., 32 (1967) 905.

130   H.J. Shine and J.T. Chamness, J. Org. Chem., 32 (1967) 901.

131   D.V. Banthorpe, E.D. Hughes and C.K. Ingold, J. Chem. Soc., (1962)
          2386.

132   D.V. Banthorpe and E.D. Hughes, J. Chem. Soc., (1962) 2402.

133   D.V. Banthorpe, J. Chem. Soc., (1962) 2407.

134   H.J. Shine and J.T. Chamness, J. Org. Chem., 28 (1963) 1232.

135   D.V. Banthorpe, E.D. Hughes and C.K. Ingold, J. Chem. Soc., (1962)
          2418.

136   D.V. Banthorpe, J. Chem. Soc., (1962) 2429.

137   J.R. Cox, Jr. and M.F. Dunn, J. Org. Chem., 37 (1972) 4415.

138   R.P. Bell, "The Proton in Chemistry", 2nd ed., Cornell University
          Press, Ithaca, N.Y., 1973.

139   C.A. Bunton and R.J. Rubin, J. Am. Chem. Soc., 98 (1976) 4236.

140   M. Eigen, Angew. Chem. Int. Ed. Engl. 3 (1964) 1.

141   W.M.J. Strachan, A. Dolenko and E. Buncel, Can. J. Chem., 47 (1969)
          3631; E. Buncel, W.M.J. Strachan, and H. Cerfontain, ibid.,
          49 (1971) 152.

142   V. Sterba and M. Vecera, Coll. Czech. Chem. Comm., 31 (1966) 3486.

143   M.D. Cohen and G.S. Hammond, J. Am. Chem. Soc., 75 (1953) 880.

144   A. Cooper, Ph.D. Thesis, London, 1966; quoted in ref. (119).

145   L.P. Hammett, Physical Organic Chemistry, 2nd edition, McGraw Hill,
          New York, 1970, p.322.

146   C.H. Rochester, "Acidity Functions", Academic Press, London, 1970,
          pp.109-115.

228

147   J.F. Bunnett and F.P. Olsen, Can. J. Chem., 44 (1966) 1917.

148   J.T. Edward, Trans. Roy. Soc. Can., (4).2, Sect. III, (1964) 313.

149   K. Yates, Acc. Chem. Res., 4 (1971) 136.

150   W.P. Jencks, "Catalysis in Chemistry and Enzymology", McGraw Hill,
      New York, N.Y. 1969.

151   D.V. Banthorpe, E.D. Hughes and C.K. Ingold, J. Chem. Soc., (1964)
      2864.

152   D.V. Banthorpe and J.G. Winter, J. Chem. Soc., Perkin II, (1972)
      868.

153   P.J. Smith, "Isotopes in Organic Chemistry", Vol. 2, Eds. E. Buncel
      and C.C. Lee, Elsevier, Amsterdam, 1976.

154   A. Fry, Chem. Soc. Revs., 1 (1972) 163.

155   K.C. Westaway, Can. J. Chem., 56 (1978) 2691.

156   C.J. Collins and N.W. Bowman, Editors, "Isotope Effects in
      Chemical Reactions", Van Nostrand, Princeton, N.J., 1971.

157   W.H. Saunders, Jr. and A.F. Cockerill, "Mechanisms in Elimination
      Reactions", Wiley, New York, 1973.

158   E. Buncel and A.N. Bourns, Can. J. Chem., 38 (1960) 2457; A.N.
      Bourns and E. Buncel, Ann. Rev. Phys. Chem., 12 (1961) 1.

159   H.J. Shine, G.N. Henderson, A. Cu, and P. Schmid, J. Am. Chem.
      Soc., 99 (1977) 3719.

160   D.H. Smith, J.R. Schwartz and G.W. Wheland, J. Am. Chem. Soc.,
      74 (1952) 2282.

161   A. Heesing and U. Schinke, Chem. Ber., 105 (1972) 3838; 110
      (1977) 3319.

162   J.R. Cox, Jr. and M.F. Dunn, J. Org. Chem., 37 (1972) 4415.

163   G.S. Hammond and J.S. Clovis, J. Org. Chem., 28 (1963) 3283.

164   M.J.S. Dewar, in "Molecular Rearrangements", Vol. 1, Ed. P. de
      Mayo, Interscience, New York, 1963, p.323; M.J.S. Dewar and
      A.P. Marchand, Ann. Rev. Phys. Chem., 16 (1965) 338.

165   H.J. Shine and J.P. Stanley, Chem. Commun., 294 (1965); J. Org.
      Chem., 32 (1967) 905.

166   U. Svanholm, K. Bechgaard, O. Hammerich and V.D. Parker,
      Tetrahedron Lett., (1972) 3675; U. Svanholm and V.D. Parker,
      Chem. Commun., (1972) 440.

167   L. Hammick and S.F. Mason, J. Chem. Soc., (1946) 638.

168   G.A. Olah, K. Dunne, D.P. Kelly and Y.K. Mo, J. Am. Chem. Soc., 94
      (1972) 7438.

169   M.E. Lupes, Rev. Roum. Chim., 17 (1972) 1253.

170   Z.J. Allan, Tetrahedron Lett., (1971) 4225; Monatsch. Chem., 106
        (1975) 429.

171   S. Winstein, B. Appel, R. Baker, and A. Diaz, Chem. Soc. Spec.
        Publ. No. 19 (1965) 109.

172   W.N. White, D. Lazdins, and H.S. White, J. Am. Chem. Soc., 86
        (1964) 1517.

173   A.K. Colter, F.F. Guzik, and S.H. Hui, J. Am. Chem. Soc., 88 (1966)
        5754.

174   C.A. Bunton and S.K. Huang, J. Am. Chem. Soc., 95 (1973) 2701.

175   A. Dolenko and E. Buncel in "The Chemistry of the Hydrazo, Azo,
        and Azoxy Groups", S. Patai, Ed., Wiley, New York, 1975.

176   E.S. Lewis, R.E. Holliday and L.D. Hartung, J. Am. Chem. Soc., 91
        (1969) 430.

177   P. Burri, H. Loewenschuss, H. Zollinger, and G.K. Zwolinski, Helv.
        Chim. Acta, 57 (1974) 395; I. Szele and H. Zollinger, ibid.,
        61 (1978) 1721.

178   E.S. Lewis and J.E. Cooper, J. Am. Chem. Soc., 84 (1962) 3847.

179   J.M. Insole and E.S. Lewis, J. Am. Chem. Soc., 85 (1963) 122.

180   A.K. Bose and I. Kugajevsky, J. Am. Chem. Soc. 88 (1966) 2325.

181   E.S. Lewis and R.E. Holliday, J. Am. Chem. Soc., 91 (1969) 426;
        E.S. Lewis and J.M. Insole, ibid., 86 (1964) 32.

182   E.S. Lewis, L.D. Hartung, and B.M. McKay, J. Am. Chem. Soc., 91
        (1969) 419.

183   R.G. Bergstrom, R.G.M. Landells, G.H. Wahl, Jr., and H. Zollinger,
        J. Am. Chem. Soc., 98 (1976) 3301.

184   I. Szele and H. Zollinger, J. Am. Chem. Soc., 100 (1978) 2811.

185   Y. Hashida, R.G.M. Landells, G.E. Lewis, I. Szele, and H. Zollinger,
        J. Am. Chem. Soc. 100 (1978) 2816.

186   S. Winstein, R. Appel, R. Baker, and A. Diaz, Chem. Soc. Spec.
        Publ. No. 19 (1965) 109 and references therein.

186a W. Maurer, I. Szele and H. Zollinger, Helv. Chim. Acta, 62 (1979)
        1079.

187   A. Cox, T.J. Kemp, D.R. Payne, M.C.R. Symons, and P.P. deMoira,
        J. Am. Chem. Soc., 100 (1978) 4779.

188   G.W. VanDine and R. Hoffmann, J. Am. Chem. Soc., 90 (1968) 3227.

189   J.D. Dill, P.v.R. Schleyer, and J.A. Pople, J. Am. Chem. Soc., 99
        (1977) 1.

190   M.A. Vincent and L. Radom, J. Am. Chem. Soc,, 100 (1978) 3306.

191   H. Zollinger, Acc. Chem. Res., 6 (1973) 335.

192   C.G. Swain, J.E. Sheats, and K.G. Harbison, J. Am. Chem. Soc., 97
          (1975) 783.

193   C.G. Swain, J.E. Sheats, D.G. Gorenstein, and K.G. Harbison, J. Am.
          Chem. Soc., 97 (1975) 791.

194   C.G. Swain, J.E. Sheats, and K.G. Harbison, J. Am. Chem. Soc., 97
          (1975) 796.

195   H. Zollinger, Angew. Chem. Int. Ed. Engl., 17 (1978) 141.

SUBJECT INDEX